奇思妙想发明史

创造现代世界

马特·特纳（英）著
萨拉·康纳（英）绘
周渝毅 译

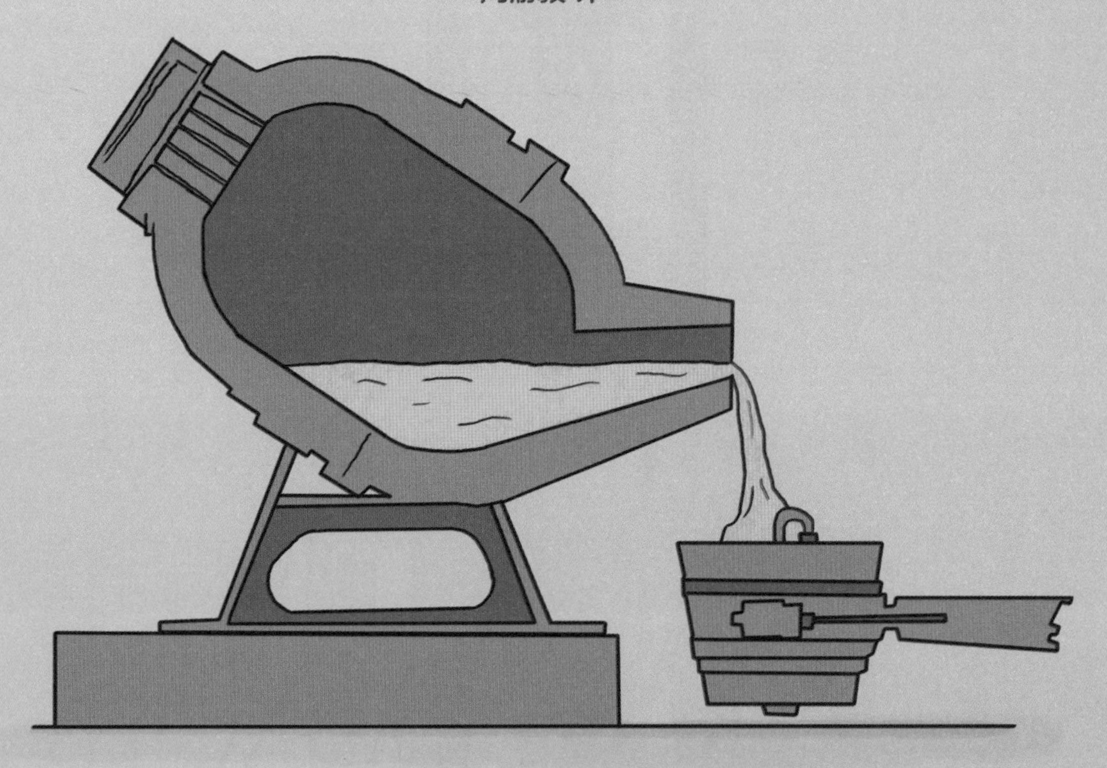

外语教学与研究出版社
FOREIGN LANGUAGE TEACHING AND RESEARCH PRESS
北京 BEIJING

工程学上无数巧妙的发明和奇迹改变了我们的世界，帮助我们改善生活品质、种植粮食、制作物品，为机械提供动力以及帮助人类强身健体。其实人类从古代就琢磨出很多聪明的做法和巧妙的工具，创新永不止步！在这本书里，你会见识到：

- 马力十足的蒸汽发动机
- 被当成“大脑营养液”售卖的汽水（猜猜是什么）
- 18世纪的“伏打电堆”（一种电池）
- 居家英雄“埃阿斯”，即1596年出现的、最早的抽水马桶
- 关于狗狗的发明（差不多可以这样说吧）

还有更多的发现等着你。你也会读到一些馊主意，比如发明地震机（发明者自己都不喜欢这个创意），还可以了解青霉素和薯片的意外诞生……

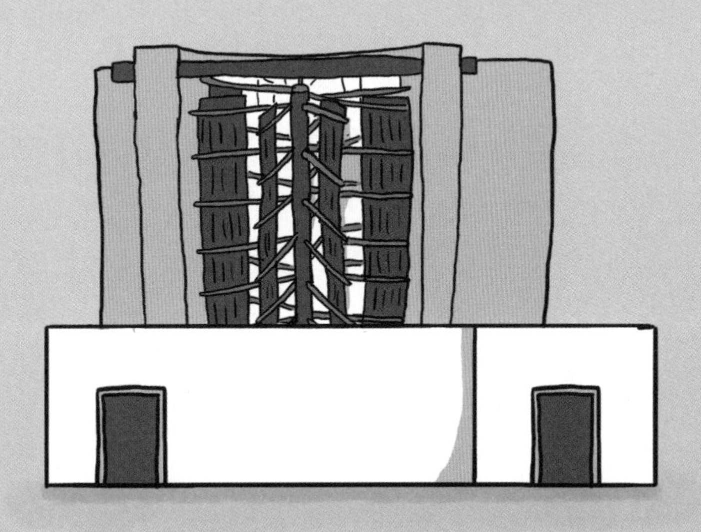

尼龙发明于20世纪30年代，人们用尼龙制作长筒袜——也用来生产降落伞和其他战争物资。

目录

导语：创新永不止步

今天我们的社会在不断地创新，每天都有新产品和新想法出现。有些发明创造很厉害，比如电动机、心脏起搏器，然而有的也不那么高明，比如托马斯·爱迪生的巨型果冻模型房屋。但是这些创意都是人类发明创造历史的一部分。人类的发明创造史可以追溯到几千年前，当时我们的祖先发明了犁、轮子、钟表等无数我们现在习以为常的日用品。这些发明创造都对社会的塑造做出了贡献：电让我们实现了即时通信，用上了方便的厨房用具，也坐上了清洁的交通工具；医学上的进步则延长了人们的平均寿命。

中世纪平地犁，九世纪

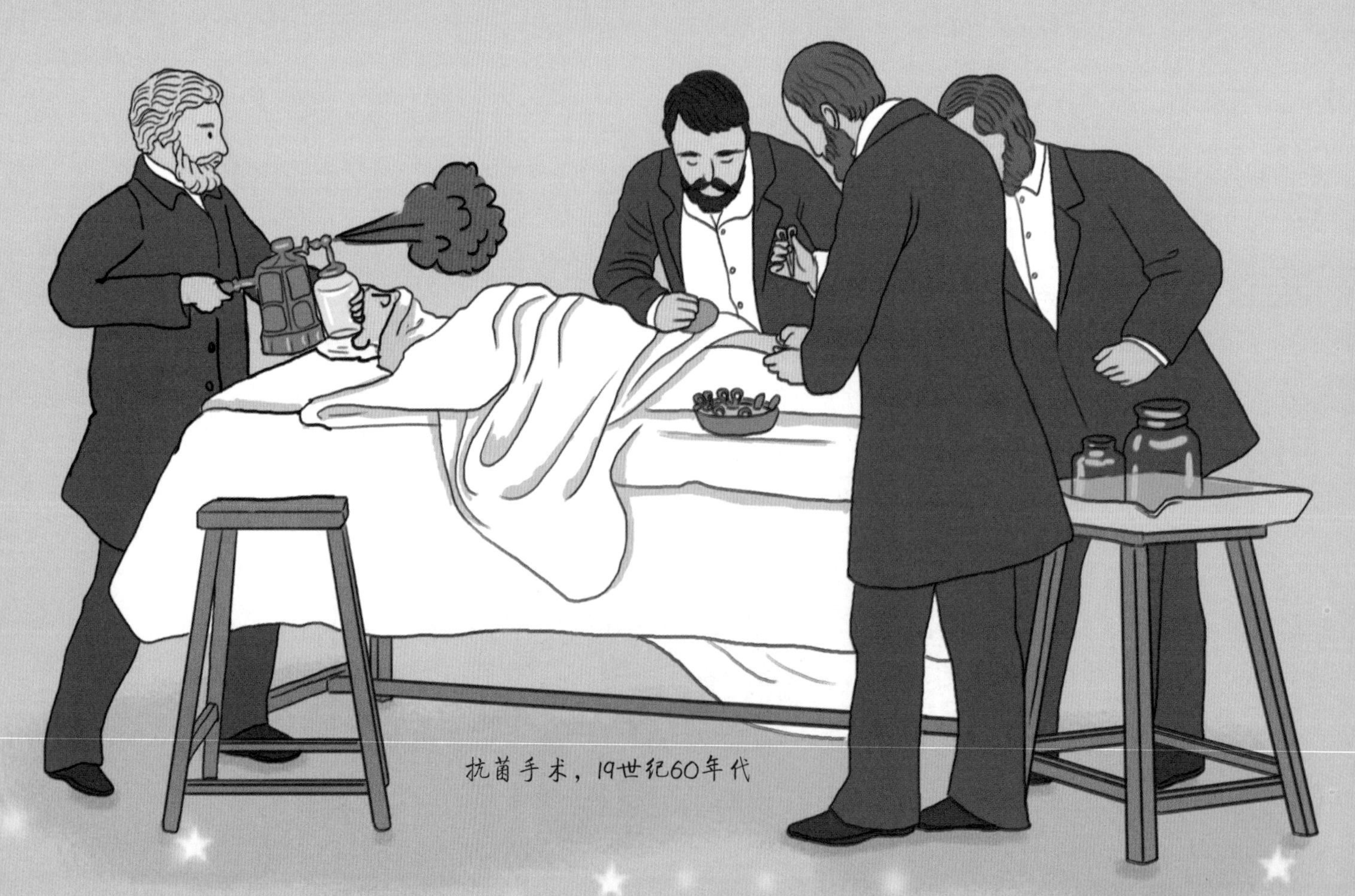
抗菌手术，19世纪60年代

我们并非总能确定这些发明是谁的主意，它们往往就是在人类逐渐走向进步的过程中出现的。但是偶尔会有想象力的飞跃，或者是偶然的发现，导致了新想法的产生。比如冰棍就是偶然发明的，当时有人把一杯苏打饮料遗忘在滴水成冰的门口，冰棍就此诞生了！

贝塞麦转炉，1856年

本书将带领我们回顾各种发明的辉煌历史——从最早的时代开始，经历工业革命的骚动，再到今天。正是这些发明创造了我们的世界。你将了解农作物的种植方式、蒸汽动力和电力的兴起、计时器的发明、各种家居用品的发明、医学上的里程碑、智能制造以及尼龙搭扣和凯芙拉等神奇材料的发明。

水力纺纱机，1767年

尼龙搭扣，1955年

最后，希望你会意识到，任何人——包括你在内——都能通过奇思妙想改变我们今天的生活方式。

农业的进步

现代农业虽然涉及高科技、大型拖拉机以及复杂的土壤科学，但是仍然会用到许多古代发明，比如犁和风车。我们的祖先利用这些工具首次开垦了荒地，并使土壤变得肥沃。

简易犁

人类大约在8,000年前驯化了牛，后来又发明了古式铧犁。上图这种简易犁最先是用木头做的，后来改用金属，能在地里犁出单行沟。

重型犁

上图所示是九世纪的重型犁，上面的犁刀会在土里先开出一道沟。犁刀后面是犁铧，能在犁壁的配合下翻土。

桔槔

古埃及人用桔槔汲水。短的一头有重物，一般是黏土或石头，使挂有水桶的那一头保持平衡，这样一个人就能很轻松地把水打上来。

独轮车

是谁发明了独轮车？没有人知道确切答案。很可能是古希腊人。但可以确定的是，大约在公元200年，古代中国人就用上了独轮车。（也是中国人最早驯化了猪和鸡。）

轻型犁

下图这种犁发明于1730年，由英格兰罗瑟勒姆的约瑟夫·福尔贾姆设计。这种犁制作起来非常简单，而且它的铁皮犁壁犁起地来非常平整。

拖拉机

现在人们可以在拖拉机后面连一个多刃犁（或其他任何农具），由拖拉机的发动机提供动力。

风车

世界上最早的风车很可能发明于八世纪的波斯，当时人们用它碾磨谷物或者汲水。这些风车“横向并排”工作，内壁水平旋转，从而让风像通过漏斗般通过叶片。

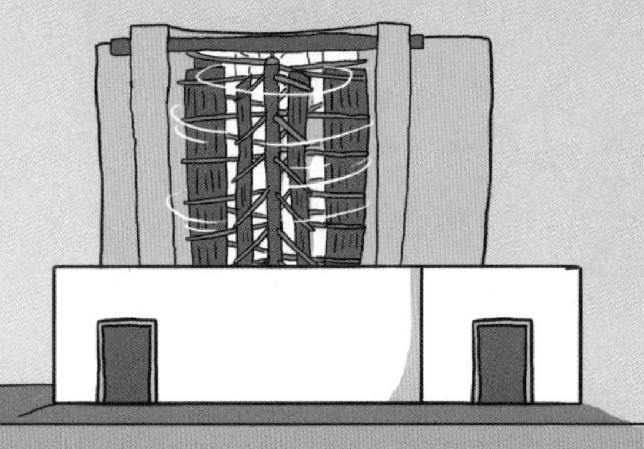

阿基米德式螺旋抽水机

阿基米德式螺旋抽水机通过转动一个螺旋形装置抽水，以希腊天才阿基米德（公元前287—公元前212年）的名字命名。但是这种抽水机可能早在阿基米德出生前几百年就已经发明出来了。

水车

大约2,000年前，罗马人在法国巴贝加尔建造了一个集16部水车为一体的“楼梯式”水车。（右图所示只是位于最底部的水车。）他们用这种水车磨坊将谷物磨成面粉。

食品与科技

你家厨房的食品储藏室里很可能有罐装和瓶装食品。而且，你得有设备让食品保持低温（用冰箱）或者高温（用保温壶）。但几百年前没有这些便利设施，连像样的炊具都没有。那么是谁发明了这些东西呢？

第一个罐头

1810年，法国人尼古拉·阿佩尔发明了世界上第一个“罐头”。他把食物放入一个玻璃罐里，用软木塞和蜡将这个罐子密封，然后把罐子放进水里煮沸。这样罐子里的食物就能保持新鲜。为了证明这个方法有效，他还保存过一整只羊！

咦，没有标签啊。希望里面装的是豌豆，不是狗粮。

易拉罐拉环

1959年，艾马尔·“厄尼”·弗拉泽在美国发明了饮料罐上的拉环。在20年的时间里，他的公司靠这个巧妙的小发明一年就能赚五亿美元。

金属罐头

上图所示是世界上最早的金属罐头，由唐金、霍尔和甘布尔公司于1813年在伦敦制造。问题是，开罐器直到1855年才发明出来！所以，图中这两人只能用锤子和凿子开罐头啦。

保温瓶

大约在1892年，苏格兰化学家詹姆斯·杜瓦发明了真空保温瓶，它可以使液体保温（低温或高温）。遗憾的是，他没有申请专利。而德国的“膳魔师”公司申请了……后来的事大家都知道了。

第一台冰箱

世界上第一台冰箱是左图这台制冰机，是奥利弗·埃文斯在美国发明的，由雅各布·珀金斯在1834年制造出来。这台机器利用极其复杂的抽水系统带走热量，从而冷却水。

富兰克林炉

1741年，聪明的美国人本·富兰克林发明了有多个开口的壁炉，这些开口能释放更多热量，而且不会产生滚滚浓烟。他谦虚地把这个炉子叫做富兰克林炉。

口香糖

富兰克林炉也可以用于烹调。上图描绘的是1848年，约翰·B. 柯蒂斯在加热一罐黏糊糊的云杉树脂，他要制作美国最早的口香糖。

可口可乐

1886年，在佐治亚州亚特兰大，美国药剂师约翰·彭伯顿开始出售用起泡饮料做的“大脑营养液”，据他称这种饮品能带来“纯粹的快乐”。它叫什么名字呢？可口可乐。

铁和蒸汽

300年前人类发明蒸汽发动机后，能够生产大量优质的钢铁，由此西方国家开始了工业革命。蒸汽本身或结合电力，能给新的工厂、汽船以及机车提供动力。

帕潘的蒸汽发动机

1707年，出生于法国的物理学家德尼·帕潘在戈特弗里德·莱布尼茨的帮助下发明了世界上最早的蒸汽发动机之一，并计划把这台机器装到一艘汽船上。但是德国威悉河上的船夫们害怕这种新科技，所以就阻止他实施这个计划。

提炼纯铁

制造早期的蒸汽发动机需要大量优质纯铁。1709年，英格兰人亚伯拉罕·达比一世想出了提炼纯铁的方法，即用焦炭（“烤”过的煤）来加热铁。纽科门的蒸汽泵使用的就是达比高炉炼出的铁。

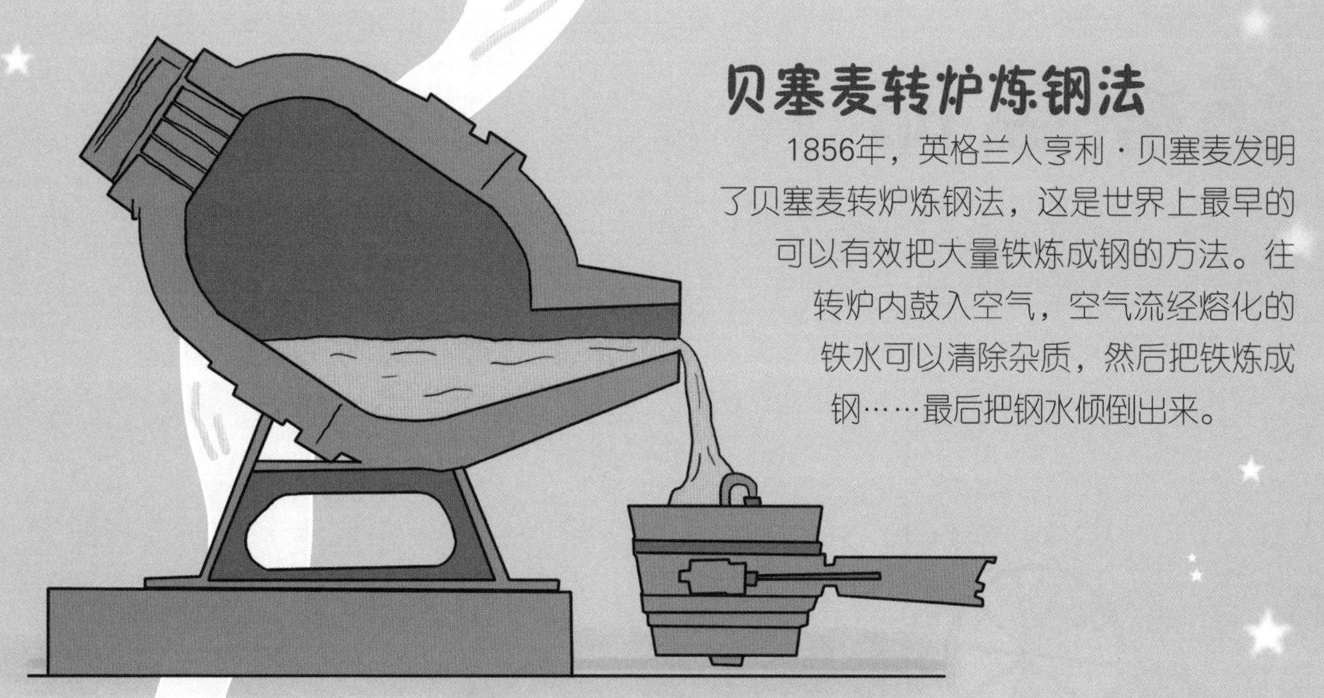

贝塞麦转炉炼钢法

1856年，英格兰人亨利·贝塞麦发明了贝塞麦转炉炼钢法，这是世界上最早的可以有效把大量铁炼成钢的方法。往转炉内鼓入空气，空气流经熔化的铁水可以清除杂质，然后把铁炼成钢……最后把钢水倾倒出来。

高压蒸汽发动机

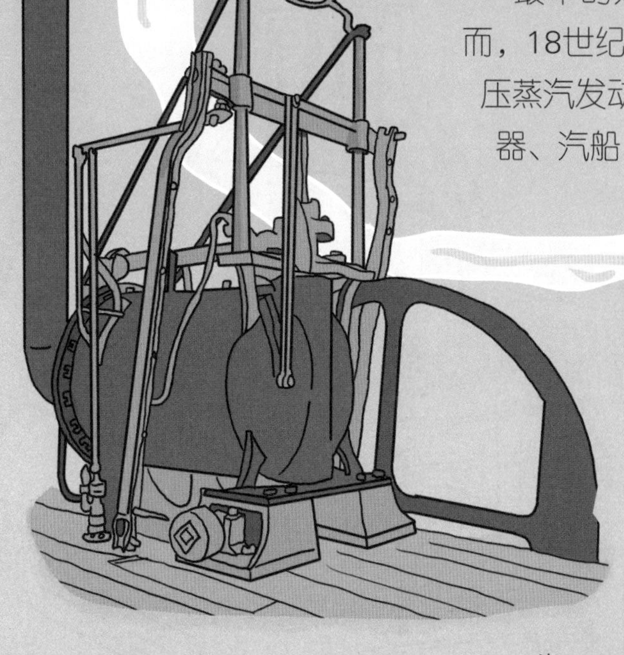

最早的蒸汽发动机使用的是低压，功率不是很大。然而，18世纪90年代英格兰人理查德·特里维西克发明的高压蒸汽发动机改变了这一状况。高压蒸汽机能驱动工业机器、汽船，还为世界上第一辆机车提供了动力。

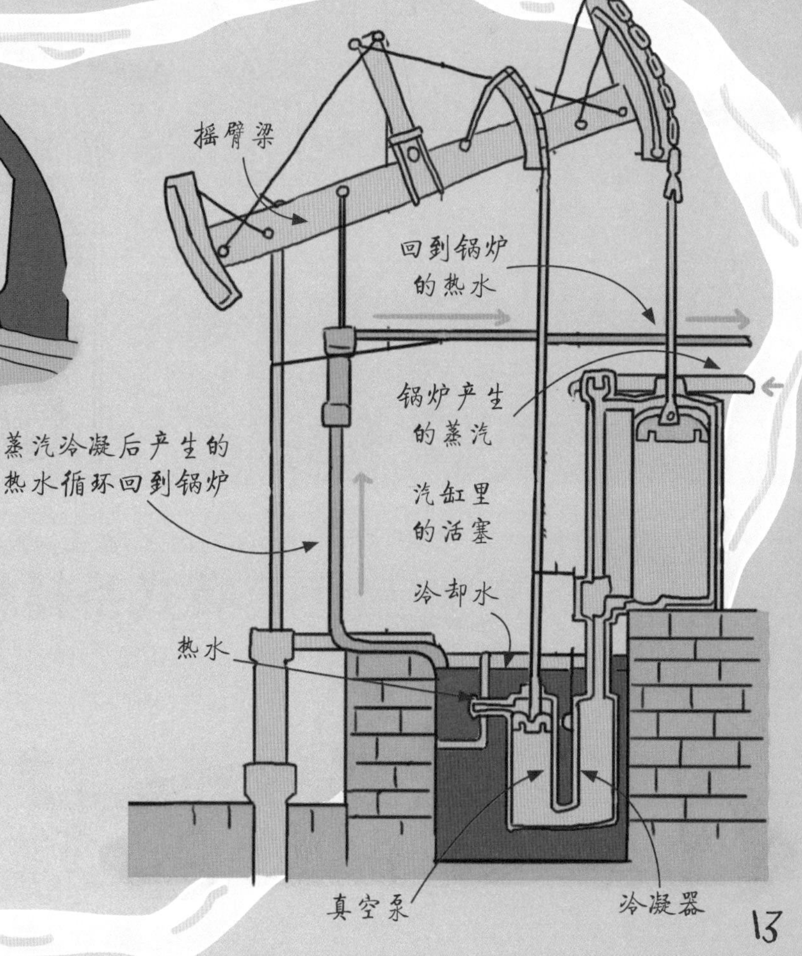

蒸汽泵

1712年，托马斯·纽科门在英格兰把帕潘的理念运用到给煤矿抽水的蒸汽发动机上。18世纪70年代，马修·博尔顿和詹姆斯·瓦特在纽科门的设计上增加了一个冷凝汽缸，大大提高了蒸汽泵（右图）的效率。

纺织革命

在工业革命中诞生的蒸汽发动机与水力共同驱动了新机器的出现，这些机器能制造出数量空前的产品，例如布匹。随着乡村人口涌入城镇新工厂工作，社会发生了巨变。

纺车

在早期，把羊毛制成纱线是一项很费时的工作。大约在公元1000年，印度发明了纺车，加快了纺纱的速度，但是纺纱工发现还是很难满足织布工对纱线的需求。

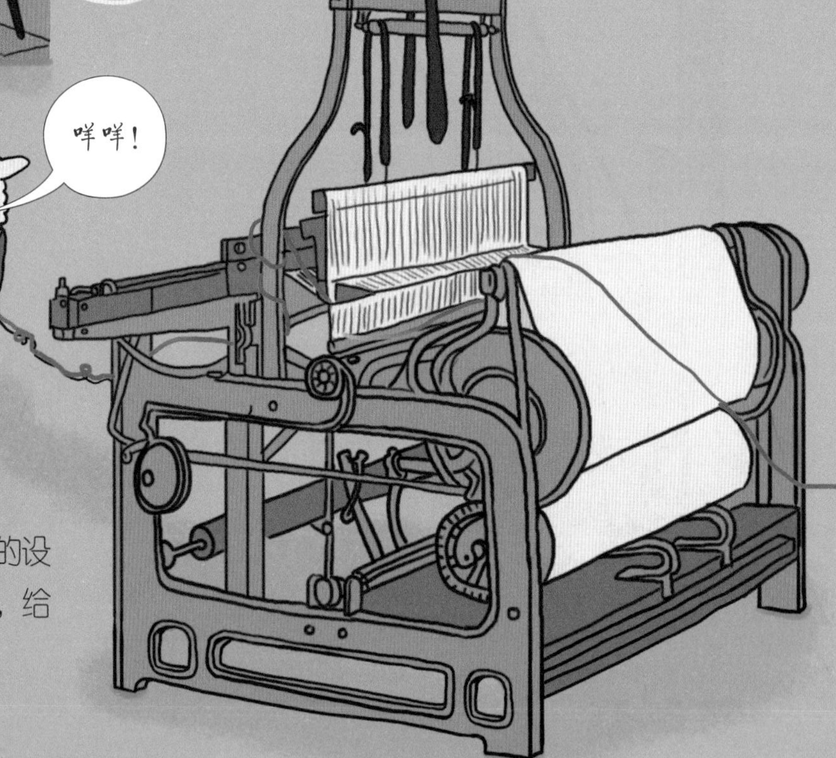

美国的纺织业

工业纺织技术传到美国要归功于弗朗西斯·洛厄尔。他在1810到1812年间访问英国时，偷偷复制了英国织布机的设计图。之后他在马萨诸塞州建造纺织厂，给工厂配备了经他改良后的新织布机。

英式条播机

机械革命也进入了农业领域。1701年，杰思罗·塔尔发明了英式条播机，能将种子以合适的生长深度放进土坑里。（以前，种子只是被撒在土壤上。）

珍妮纺纱机

1764年，英格兰人詹姆斯·哈格里夫斯发明了珍妮纺纱机，这种纺纱机一次可以纺好几个纱锭。一台机器顶得上八个、十个甚至100个纺纱工。

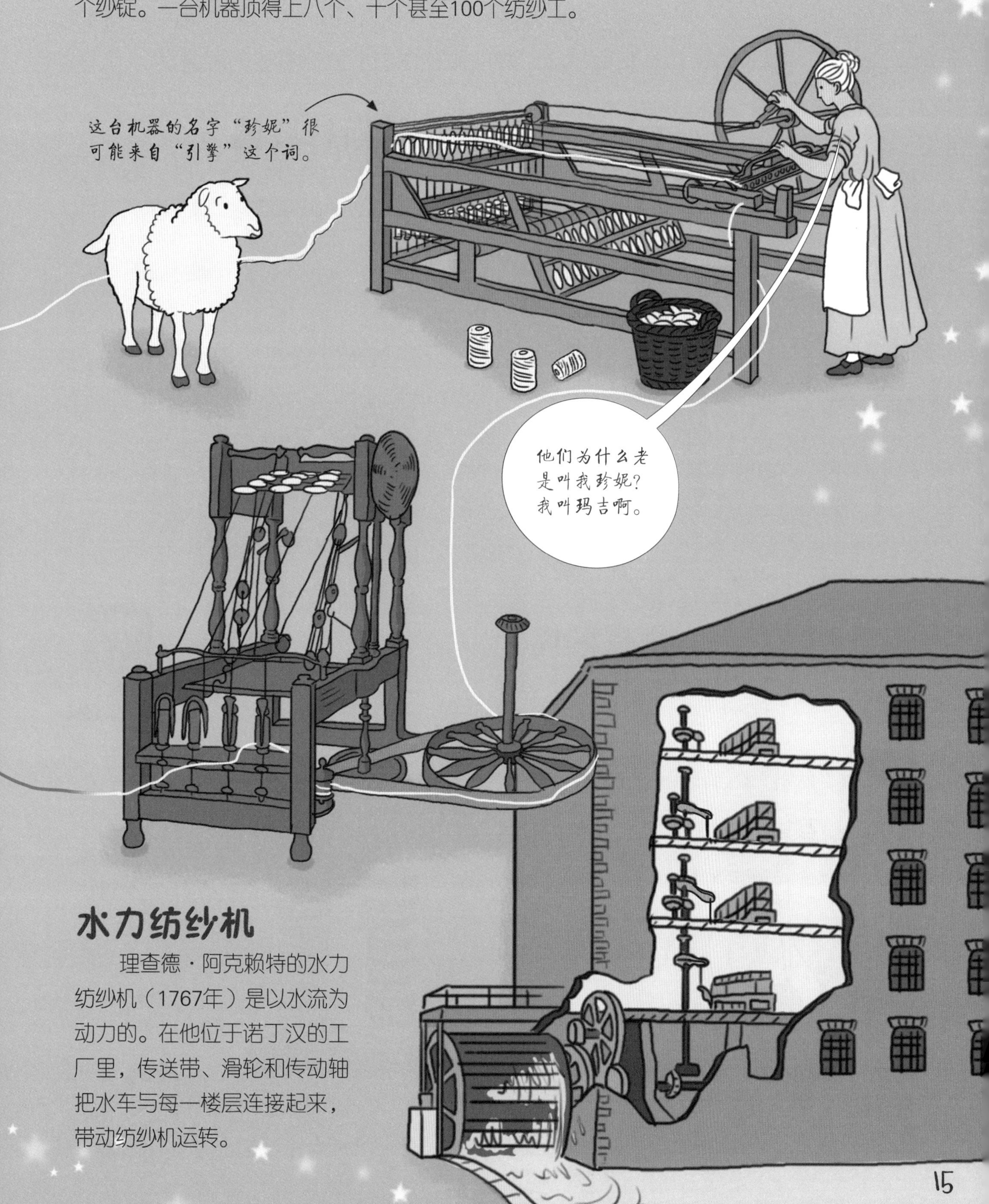

这台机器的名字“珍妮”很可能来自“引擎”这个词。
他们为什么老是叫我珍妮？我叫玛吉啊。

水力纺纱机

理查德·阿克赖特的水力纺纱机（1767年）是以水流为动力的。在他位于诺丁汉的工厂里，传送带、滑轮和传动轴把水车与每一楼层连接起来，带动纺纱机运转。

人类对电的探索

电的历史非常悠久。早期人类知道发电鱼、闪电和静电（比如抚摸猫时触电的感觉），但是储存电以及将电当成一种能源使用则要追溯到350年前。

莱顿瓶

右图是一个莱顿瓶，是一种非常原始的电池，也是一种电容器，能把电储存起来。莱顿瓶是德国的埃瓦尔德·冯·克莱斯特和荷兰莱顿的彼得·范米森布鲁克在1745年左右发明的。

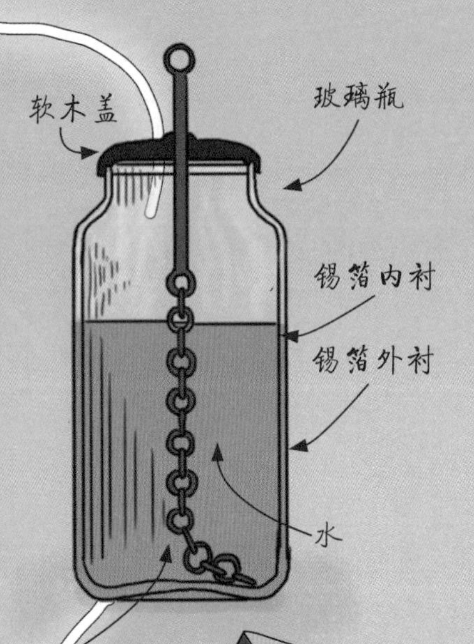

捕捉闪电

据说在1752年，不知道是本·富兰克林还是他的儿子把风筝放上天去“捕捉”闪电，并把闪电储存在莱顿瓶中。但很可能他们两人都没干过这事。不过富兰克林意识到闪电就相当于电。后来他发明了避雷针，还造出了battery（电池）这个词。

哇，爸爸，你好聪明啊!

特斯拉线圈

塞尔维亚人尼古拉·特斯拉（1856—1943）是发明无线电、霓虹灯、交流电、X射线、无线网络、甚至“死亡射线”（幸亏他没造出来）的先驱。他发明的特斯拉线圈是一种变压器。

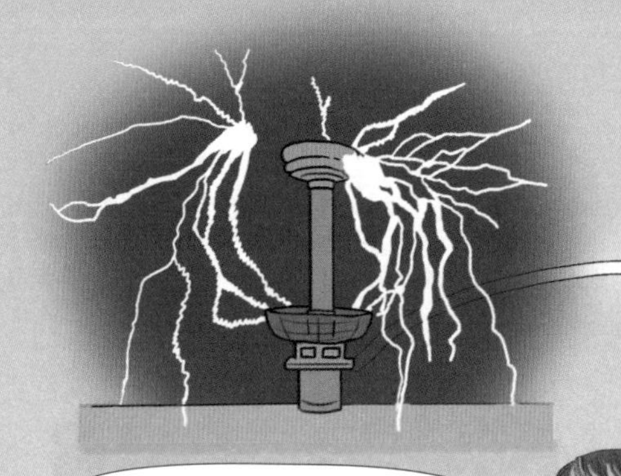

你可以把我接到电源上!

伏打电堆

18世纪90年代，意大利科学家亚历山德罗·伏打发明了“伏打电堆”，即一种电池。他的同胞路易吉·加尔瓦尼之前就发现，已死青蛙的腿通上电后会产生抽搐反应，这说明神经信号就是带电的“信息”。

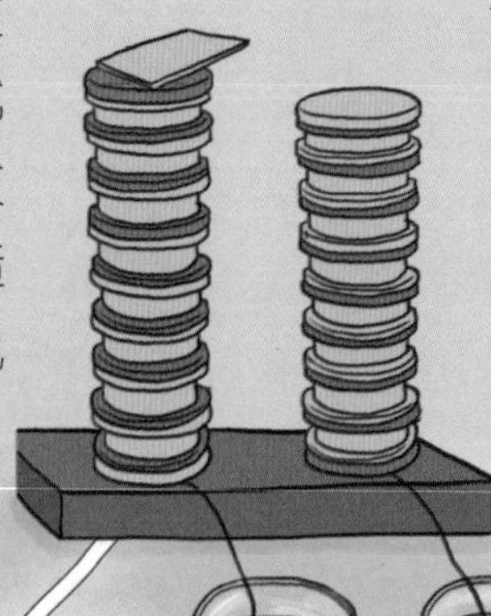

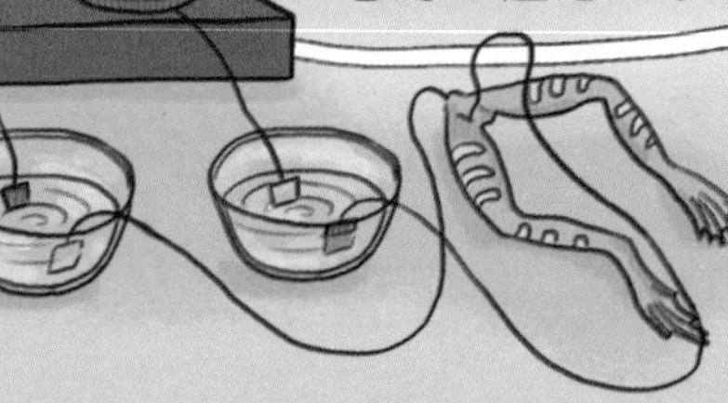

“电气化”发明家

法国人古斯塔夫·特鲁夫（1839—1902）设计了右图这款亮闪闪的珠宝。他的电气化发明还包括独木舟、舷外发动机、牙医电钻（听着就疼）、金属探测器、飞船、剃须刀、缝纫机、麦克风、电报机、来复枪等等。

我身上亮闪闪的东西可能是“电灯”，但好沉啊！

避雷针

19世纪电气化产品受到狂热追捧。右图这个男子打着伞，伞上装有避雷针，防止被雷击。

水力发电

世界上第一座用水力发电的房子位于英格兰克拉格塞德。发电机不仅能为房子提供照明，还能给附近的农舍提供电力。

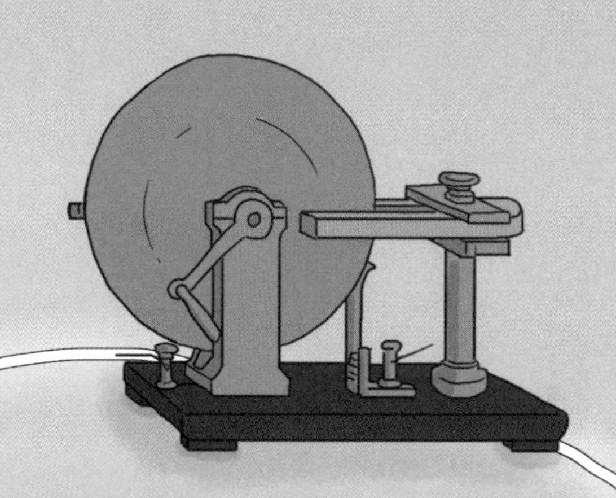

电动机

1869年，泽诺布·格喇姆发明了发电机，但这台机器不太完善。后来他（意外地）发现这台机器连上电源后会自己旋转。由此他发明了现代电动机。

法拉第圆盘

英格兰科学家迈克尔·法拉第将电和磁联结在一起，迈出了关键性的一步。1831年，他发明了早期的发电机——法拉第圆盘。这种圆盘是通过让磁场里的金属飞轮快速旋转来发电的。

哎呀！不过还是很厉害啊！

计时器的发明

人类利用各种方法来计算时间的流逝。古人看着太阳每天东升西落，便聪明地用它来计时。当然，日晷无法在夜间使用，因此发条装置是接下来的一大进步。

3D日晷

下图这个3D日晷是公元前250年由希腊天文学家阿利斯塔克发明的，叫做仰仪。立杆在半球内部投下影子来显示时间。（说到这里要提一句，阿利斯塔克曾用这个方法计算地球的大小以及地球与月球之间的距离。）

水钟

古埃及人依靠水钟（也叫滴漏计时器）来计时：即让水从一个容器滴入另一个容器。（古埃及人辩论发言也用它来计时。水滴完了，发言的人也该闭嘴了！）

超级计时器

亚历山德里亚的克特西比乌斯于公元前三世纪设计了一座水钟，据说它是此后1,800年中最精确的计时器。

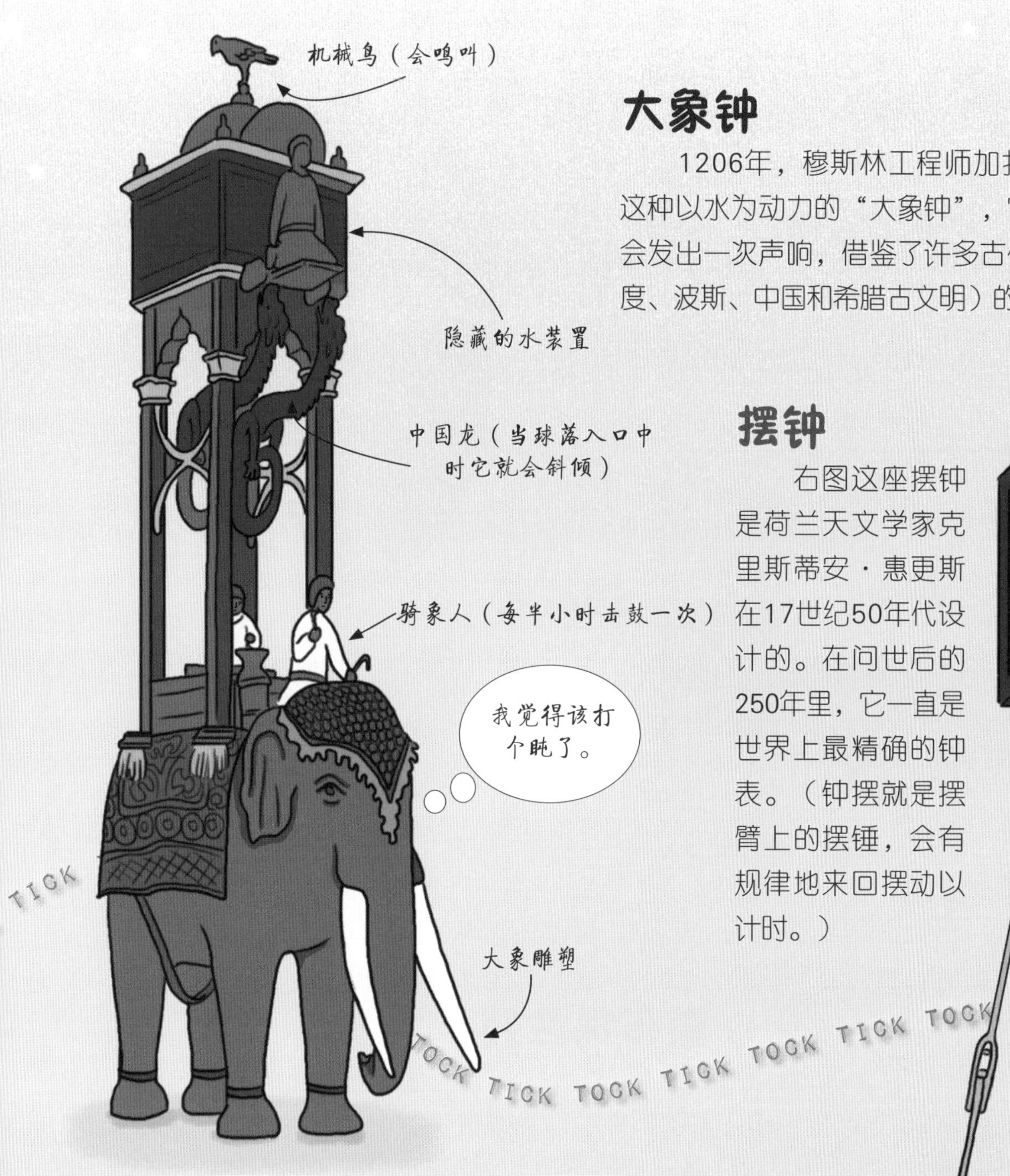

大象钟

1206年，穆斯林工程师加扎利发明了左图这种以水为动力的“大象钟”，它每隔半小时就会发出一次声响，借鉴了许多古代文明（包括印度、波斯、中国和希腊古文明）的技术。

摆钟

右图这座摆钟是荷兰天文学家克里斯蒂安·惠更斯在17世纪50年代设计的。在问世后的250年里，它一直是世界上最精确的钟表。（钟摆就是摆臂上的摆锤，会有规律地来回摆动以计时。）

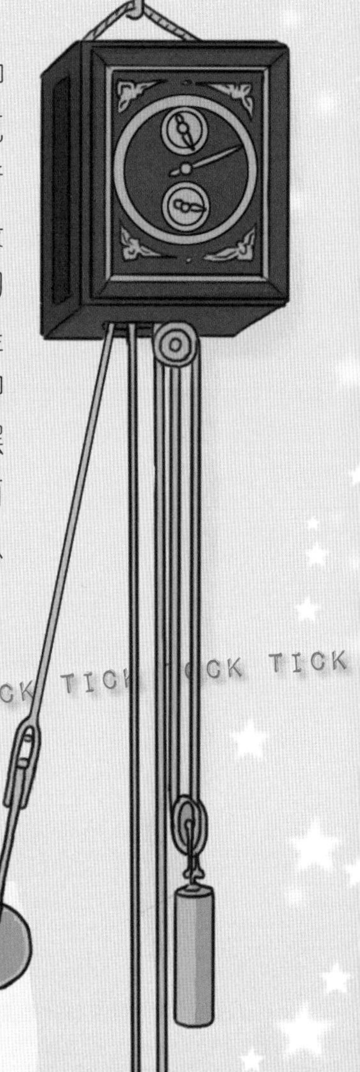

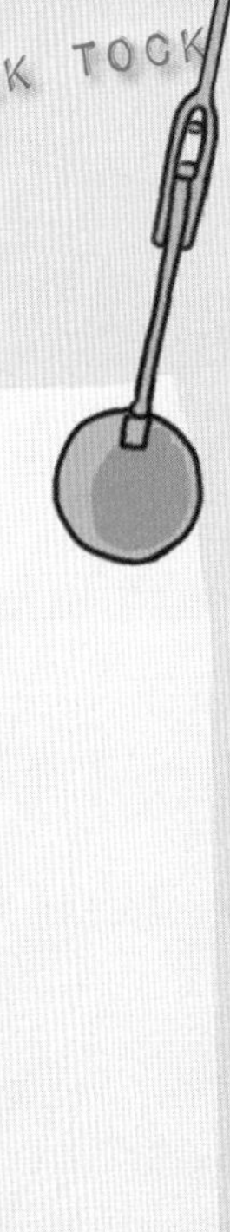

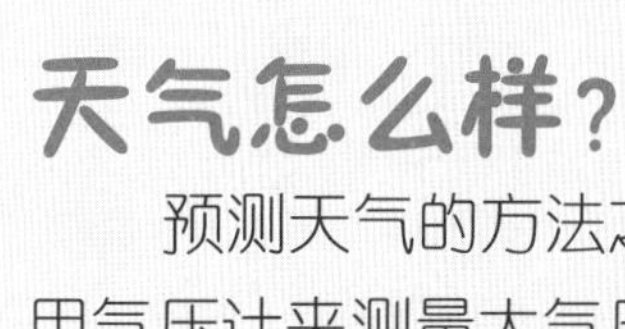

天气怎么样？

预测天气的方法之一是用气压计来测量大气压。左图这个早期的气压计诞生于1643年，是意大利人埃万杰利斯塔·托里拆利的发明。它会显示水银（一种液体金属）的高度变化。

巧妙的家居用品

你在家里和学校四处转转就会发现一些基本的材料，其中一些材料，比如玻璃、水泥、砖块，其历史可追溯到古代文明时期，但我们很难确认发明者是谁。不过可以确定的是，那些巧妙的装置，比如锁、门铃、电梯和抽水马桶，都来自某些人的灵感。

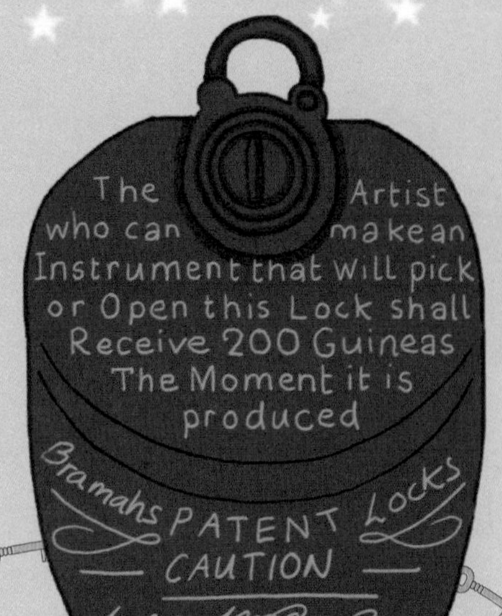

The Artist
who can make an
Instrument that will pick
or Open this Lock shall
Receive 200 Guineas
The Moment it is
produced
Bramahs PATENT LOCKS
CAUTION

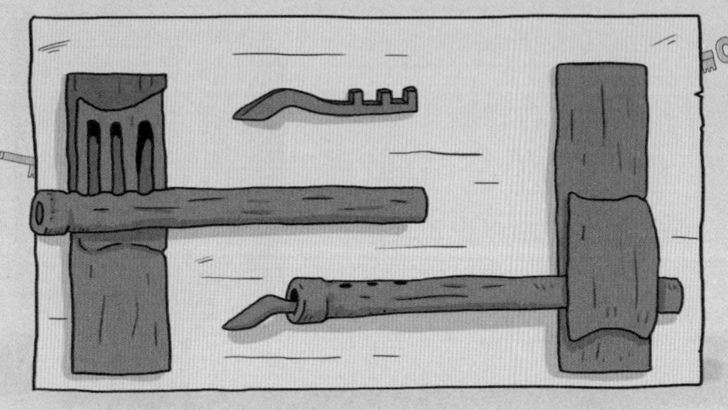

销子锁

2,000多年前，美索不达米亚人使用的是以钥匙开启的木质门锁。上图是古埃及人使用的一把销子锁。

挑战锁

安全性高的锁最早产生于1784年，是当时英国工程师约瑟夫·布喇马设计的“挑战”锁。他许诺说，有人如果（不用钥匙）打开这把锁，就能得到一大笔钱。67年后，这把锁才被人打开！当然，这时老约瑟夫早就去世了。

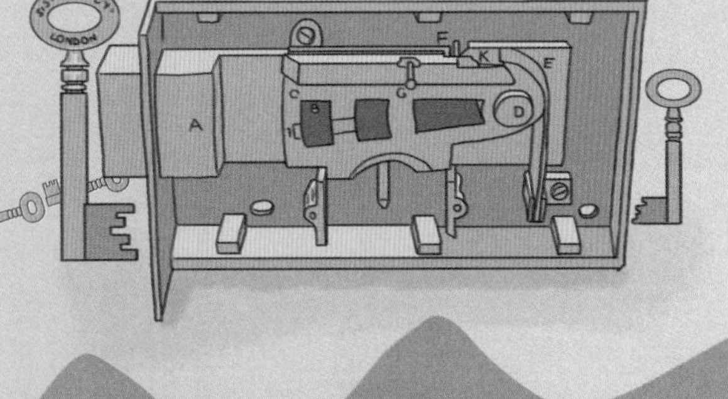

查布保险锁

1818年，英格兰锁匠杰里迈亚·查布发明了“探测锁”。如果钥匙用得不对，锁就会卡住，只有用一把特殊的备用钥匙才能打开（如左图所示）。

电门铃

1831年，美国科学家约瑟夫·亨利发明了一种能够通过一根电线在室内发出铃声的门铃。是的，拜他所赐才有了那些按了门铃就跑的熊孩子们。（约瑟夫门铃里的继电器后来被塞缪尔·莫尔斯用在了莫尔斯电码的电键里）。

这门铃真是烦死我了。

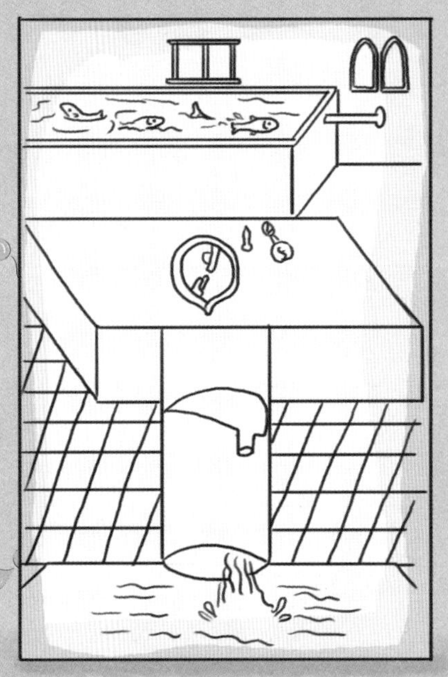

马桶

如果说家里有一样东西是我们每天都要用的，那必定是马桶了。马桶也是一项历史悠久的发明。罗马人早就开始使用带有排水系统的马桶来冲走粪便，而且他们还把海绵绑在棍子上擦屁股。

抽水马桶

1596年，英格兰作家约翰·哈林顿爵士为“埃阿斯”——世界上第一个真正的抽水马桶——画了上面的设计图。（在他那个时代，抽水马桶被叫做“茅房”。）不过，他很可能并没有在马桶水箱里养鱼！

能持久使用的马桶

锁匠约瑟夫·布喇马也设计了抽水马桶。他为自己在伦敦的锁匠铺修建了抽水马桶，有一些至今仍在使用。

安全电梯

早期的电梯十分危险，如果提升缆索断裂，电梯就会坠落到地上，造成坐电梯的人死亡。因此，美国人伊莱沙·奥蒂斯发明了带有紧急刹车装置的安全电梯，并在1854年举办的纽约世界博览会上将其展出。电梯的缆索被割断后，电梯里的伊莱沙只下降了几英寸，刹车装置就让电梯停下了。他的安全电梯很成功！

神奇的材料

也许你没听说过材料科学，但这可是一个大领域。发明家们在不断琢磨如何改善制作东西的材料，例如使材料更便宜好用或者更环保等。欢迎走进“材料”历史和科学！

“麦金托什”雨衣

“麦金托什”雨衣是苏格兰人查尔斯·麦金托什于19世纪20年代发明的。这种雨衣能防水，因为麦金托什在两层布料之间加入了防水橡胶。

我的衣服一点儿也没湿（所以请不要开“湿嗒嗒”的玩笑！）。

尼龙搭扣

尼龙搭扣是一只狗发明的，差不多可以这么说吧。1941年，乔治·德梅斯特拉尔在瑞士散步时，发现一种植物弯钩状的毛刺粘在他的狗身上。这给了他启发，他想发明一种黏性条状扣件，一面是“毛皮”似的东西，另一面是微小的弯钩。1955年，尼龙搭扣问世了。

没错，这东西是我发明的！

杜邦公司的发明

美国杜邦公司的员工善于发明新材料，以下介绍了他们最著名的发明。

尼龙

战争时期用的降落伞是用尼龙制作的。1935年，华莱士·卡罗瑟斯发明了尼龙，他当时是用尼龙生产女式长筒袜。

氟利昂

多年以来，氟利昂或以气体形式被存入小型喷雾罐中，或被用为冰箱的冷却剂。实际上杜邦公司并没有发明氟利昂，只是在1930年时开始销售氟利昂。今天，氟利昂用得并不多了，因为它会造成全球变暖。

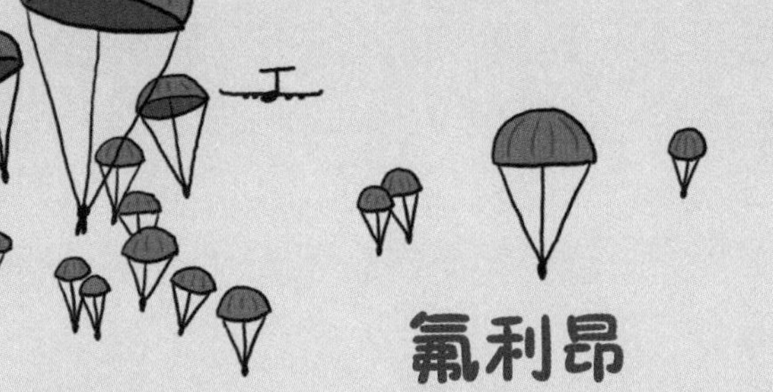

特氟隆

特氟隆是一种不粘锅涂层，方便你爸妈抛煎蛋。相比聚四氟乙烯来说，特氟隆这个名字更简单，它是罗伊·普伦基特在1938年发现的。

戈尔特斯面料

1969年，鲍伯·戈尔把一条特氟隆拉长十倍后，发现它变成了一种既防水又透气的新型面料。人们把它叫做戈尔特斯，用它来制造雨衣和步行靴。

凯芙拉

左图这只警犬穿着一件用凯芙拉制作的防弹服。凯芙拉是一种特别坚韧的材料，1965年由化学家斯蒂芬妮·克沃勒克发明。

医学发明的故事

一本书再厚，也写不完所有关于医学发现和发明的趣事。以下只列举了几个医学史上的里程碑式事件——从研究人体结构、缓解痛苦、预防疾病到创造良好的卫生条件。

研究解剖学

荷兰人安德里斯·范韦泽尔（1514—1564）是世界上最早通过解剖尸体来研究人体构造的科学家之一，人们一般叫他维萨里。他出版了许多书，书里全是各种各样的解剖图。

我现在能动了吗？我在这儿坐了12年了！

最早的疫苗

1796年，英格兰科学家爱德华·詹纳给一个小男孩接种牛痘，牛痘来自一位挤奶女工水疱里的脓汁。小男孩接种后没有得天花，天花在当时可是一种致命的疾病。多亏了詹纳在疫苗领域的开创性工作，天花才得以被消灭。

听诊器

1816年，法国医生勒内·拉埃内克在给一位妇女听心跳时发明了听诊器。他把纸卷成筒状，把耳朵贴近其中一端，然后听到了清晰的心跳声。接下来他做了一个木管式听诊器。今天的听诊器看上去和早先的样子大相径庭，但它们的工作原理是一样的。

假牙

假牙在几百年前就发明出来了，地点可能是在意大利。但是美国总统乔治·华盛顿（1732—1799）的假牙是最有名的。制作假牙的材料来自河马、奶牛和人的牙齿，可能还有象牙！

巴氏消毒法

法国化学家路易·巴斯德对细菌、疾病和疫苗的研究已经挽救了无数人的生命。1862年，他发明了（以自己名字命名的）巴氏消毒法，通过加热牛奶和啤酒来杀死细菌。左图中，巴斯德正看着他的同事埃米尔·鲁给一个男孩打狂犬疫苗。

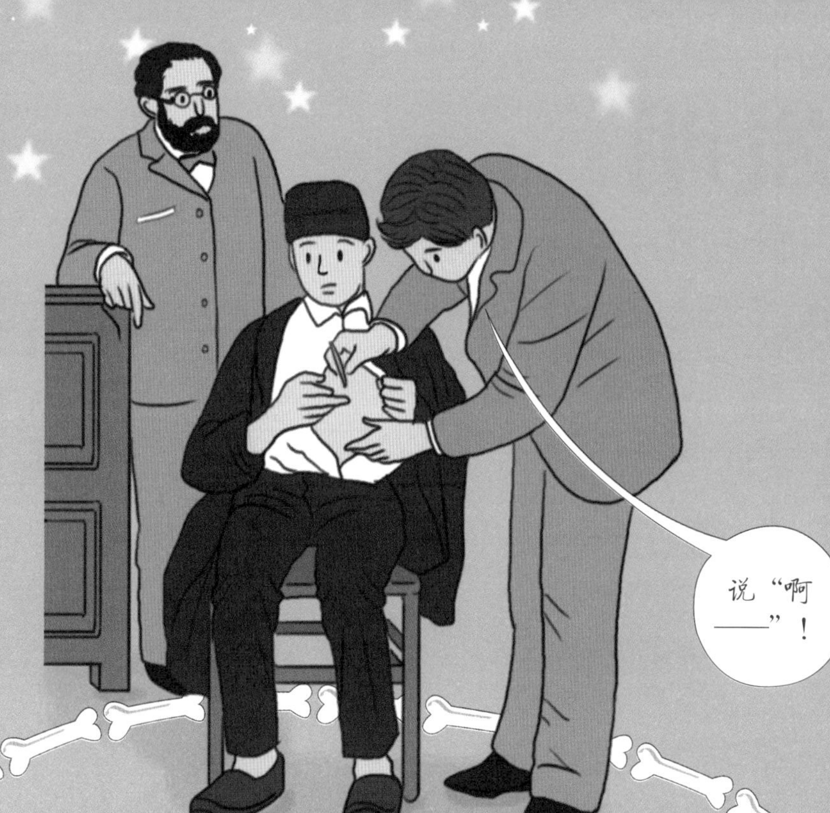

医疗卫生

以前外科医生做手术前不洗手，现在看来简直难以相信。1847年，匈牙利医生伊格纳茨·塞麦尔维斯认为，这种不卫生的做法可能会传播细菌，导致患者死亡，于是推荐大家洗手。他因此多多少少“发明”了医疗卫生这一理念。

抗菌手术

从1867年起，受巴斯德和塞麦尔维斯的启发，英国外科医生约瑟夫·利斯特通过给患者喷洒石炭酸——一种抗菌剂或消毒剂——来“清理”手术。这能杀菌并预防感染。

工厂里的发明

发明出好产品是一回事，而想出好方法来制造这些产品则是另一回事。能批量生产的工厂历史悠久，但是现在我们意识到大型工厂会造成浪费和污染，因此开始进行智能化生产。

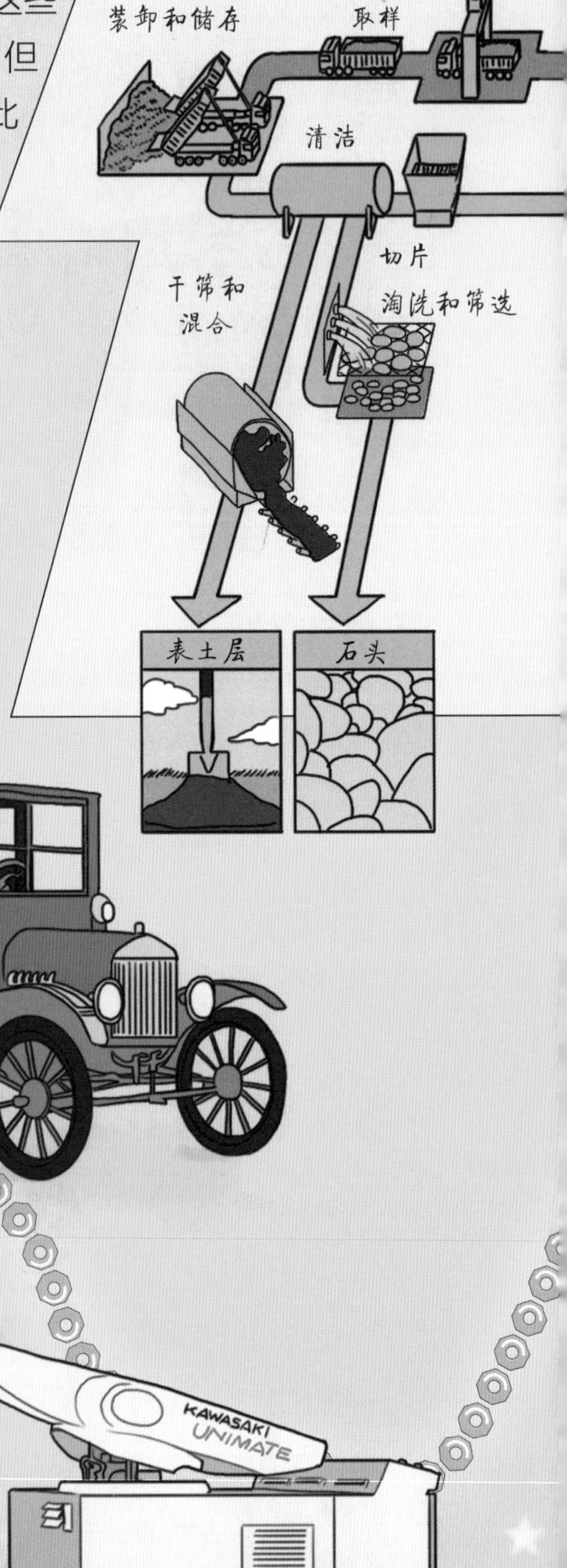

批量生产

古腓尼基人很早就开始批量生产。他们的船只就是用预制件组装而成的，就跟组装从宜家买来的书架或模型套装差不多。

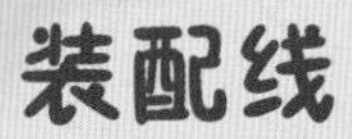

装配线

大约900年前，威尼斯人就能批量生产船只。他们发明了装配线，能在生产过程中将产品从一个厂区传送到另一个厂区。1913年，美国汽车制造商亨利·福特借鉴了这一方法来大批量组装他的T型汽车。

机器人

现代汽车工厂使用计算机程序控制机器人，来代替工人做那些特别枯燥、肮脏、困难或危险的工作。第一个工业机器人就是右图所示的“尤尼梅特”机械手臂，是乔治·迪沃尔于20世纪50年代在美国设计的。

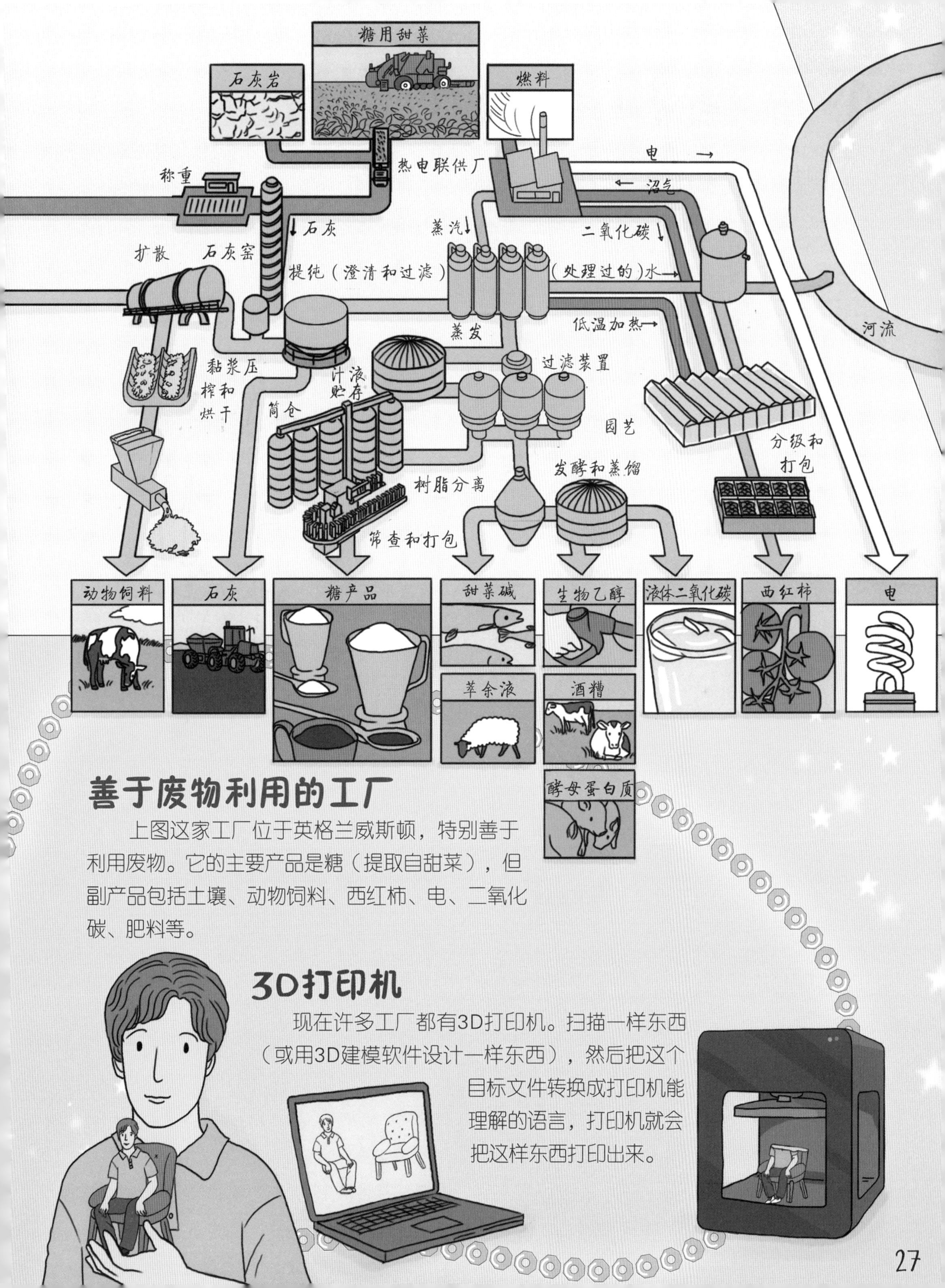

善于废物利用的工厂

上图这家工厂位于英格兰威斯顿，特别善于利用废物。它的主要产品是糖（提取自甜菜），但副产品包括土壤、动物饲料、西红柿、电、二氧化碳、肥料等。

3D打印机

现在许多工厂都有3D打印机。扫描一样东西（或用3D建模软件设计一样东西），然后把这个目标文件转换成打印机能理解的语言，打印机就会把这样东西打印出来。

不靠谱和危险的发明

任何人，无论是著名发明家还是普通人，都能提出制作小玩意的新想法。但这并不意味着每个新想法都是好点子，下面就是一些例子！

吓人的洋娃娃

灯泡、唱机以及电影摄影机的发明者托马斯·爱迪生有时也会出一些馊主意。1890年，他给孩子们设计了右图这种巨型洋娃娃。每个洋娃娃体内都有一个能“唱”儿歌的留声机。但是这种洋娃娃很贵，而且还会发出可怕的声音，顾客吓坏了，就把洋娃娃退回了商店。爱迪生后来把这些洋娃娃叫做“小怪物”。

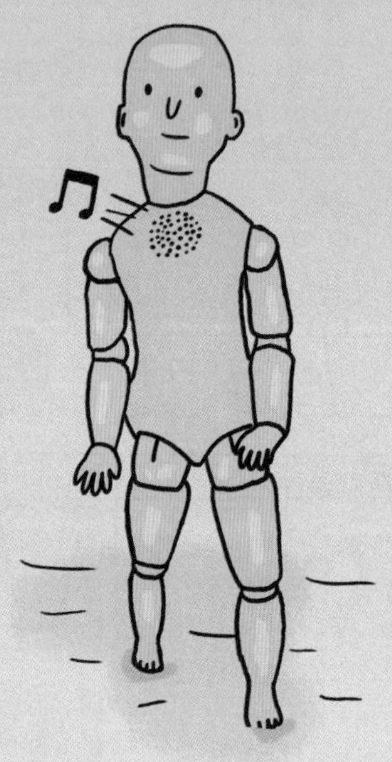

果冻模型房屋

1899年时穷人缺少住房。爱迪生想像制作巨型果冻一样通过模具建造廉价的水泥房子，从而帮助穷人。只是这些房子的造价并不便宜：比普通住房造价高40倍！今天，爱迪生的水泥房子只有一座保存了下来，位于新泽西州的蒙特克莱。

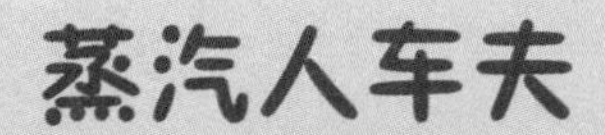

蒸汽人车夫

右图这个“蒸汽人”是扎多克·戴德里克在1868年设计的，用于拉车。它被设计成男人的样子，以免吓坏街上的马。这一设计其实没什么用，但看上去很厉害。对了，蒸汽排气管巧妙地隐藏在它的大礼帽里。

地震制造器

电气天才尼古拉·特斯拉也有失手的时候。他曾经发明了一台制造地震的机器，（理论上）能把建筑物震倒。特斯拉认为这台机器太危险了，于是用锤子把它砸烂了，从此再也未提过此事。

婴儿装甲车

第二次世界大战前夕，英国面临被敌军毒气袭击的威胁，于是一位叫米尔斯的先生发明了右图这种防空袭及毒气的婴儿推车。车顶部开了一个窗户，这样妈妈能看到宝宝是不是还在呼吸。

无心的发明

冰棍是阴差阳错发明出来的。以下还有一些皆大欢喜的意外发明。

青霉素

1928年，苏格兰生物学家亚历山大·弗莱明发现了一种能消灭疾病的药物——青霉素。因为之前他的鼻涕滴在了实验室一个标本培养皿上，然后培养皿上发霉了。

上帝保佑你，伙计。

弹簧玩具

1943年，美国海军工程师理查德·詹姆斯发明了弹簧玩具。它本来要被用作军舰里的悬挂弹簧，但詹姆斯发现，如果让弹簧倾斜，它就会自动弹回去。

一看就知道，他发明了新的超级胶水。

强力胶

第二次世界大战期间，美国化学家哈里·库弗尝试用氰基丙烯酸酯——一种透明的树脂——制作射击瞄准具和飞机座舱盖。虽然他失败了，但做出来的东西真的很黏！今天我们把这种东西叫做接触型黏合剂，或“强力胶”。

炸薯片

1853年，纽约一位食客把他买的法式炸薯条退了回去，抱怨薯条切得太厚。厨师乔治·克拉姆很生气，就把这些薯条切得非常薄，烤得非常焦，还在上面刷了一层盐，想把它做成黑暗料理。但那位食客很爱吃！由此乔治发明了薯片。

植入式心脏起搏器

心脏起搏器是一种植入体内的电子仪器，能帮助人们稳定心率。1956年，美国电气工程师威尔逊·格雷特巴奇意外发明了心脏起搏器，因为他在做一个记录装置时用错了部件。起搏器于1960年首次投入使用，从那时起，这一发明延长了无数人的生命。

索引

作者简介

马特·特纳出生于英国，20世纪80年代毕业于拉夫伯勒大学艺术学院，毕业后一直担任图片研究员、编辑和作者。他的书题材广泛，涉及博物学、地球科学和铁路等，并为百科全书和分册出版的丛书写过数百篇文章，从大象到抽象艺术无所不包。他现在和家人住在新西兰奥克兰附近，他还是当地海岸警卫队的志愿者，平时也涉猎工艺品的制作。

绘者简介

萨拉·康纳生活在英格兰美丽的乡村，栖身于一座可爱的小木屋里，有几只狗和一只猫做伴。她平时用画笔描绘身边的世界。她总能从大自然中获得灵感，这对她的很多作品都有影响。萨拉之前用钢笔和颜料画插图，但近些年，她开始在电脑上绘画，因为这更适应今天的产业发展。然而，她还是喜欢时不时拿出水彩颜料来，画一画自己花园里的鲜花！

Countless clever discoveries and wonders of engineering have changed the world and helped us to improve our lives, grow food, make things, power machinery and keep healthy. Smart practices and gadgets began appearing in ancient times – and never stopped! Spot the

- mighty steam engine
- fizzy drink sold as a 'brain tonic' (guess what that is)
- 18th-century 'voltaic pile' (that's a battery)
- 'Ajax', a domestic hero aka the first flushing loo, back in 1596
- doggy invention (well, almost)

and so much more, You'll also find daft ideas, like an earthquake machine (even the inventor didn't dig that one), and happy accidents that gave us penicillin... and crisps...

Nylon was invented in the 1930s to make stockings – *but* was also used for parachutes and other war supplies.

CONTENTS

MAKING PROGRESS

The society we live in today is in a constant state of reinvention. Every day, new products and ideas come to market. Some are excellent, like the electric motor or the heart pacemaker, while others are not so good – such as Thomas Edison's giant jelly-mould houses. But these ideas all add to a history of innovation that dates back thousands of years to when our ancestors first came up with the plough, the wheel, the clock and the countless other everyday items we take for granted. And they all shape society: electricity gives us instant communication, convenient kitchen gadgets and clean transport; medical advances have added years to the average life span.

Medieval mouldboard plough, ninth century

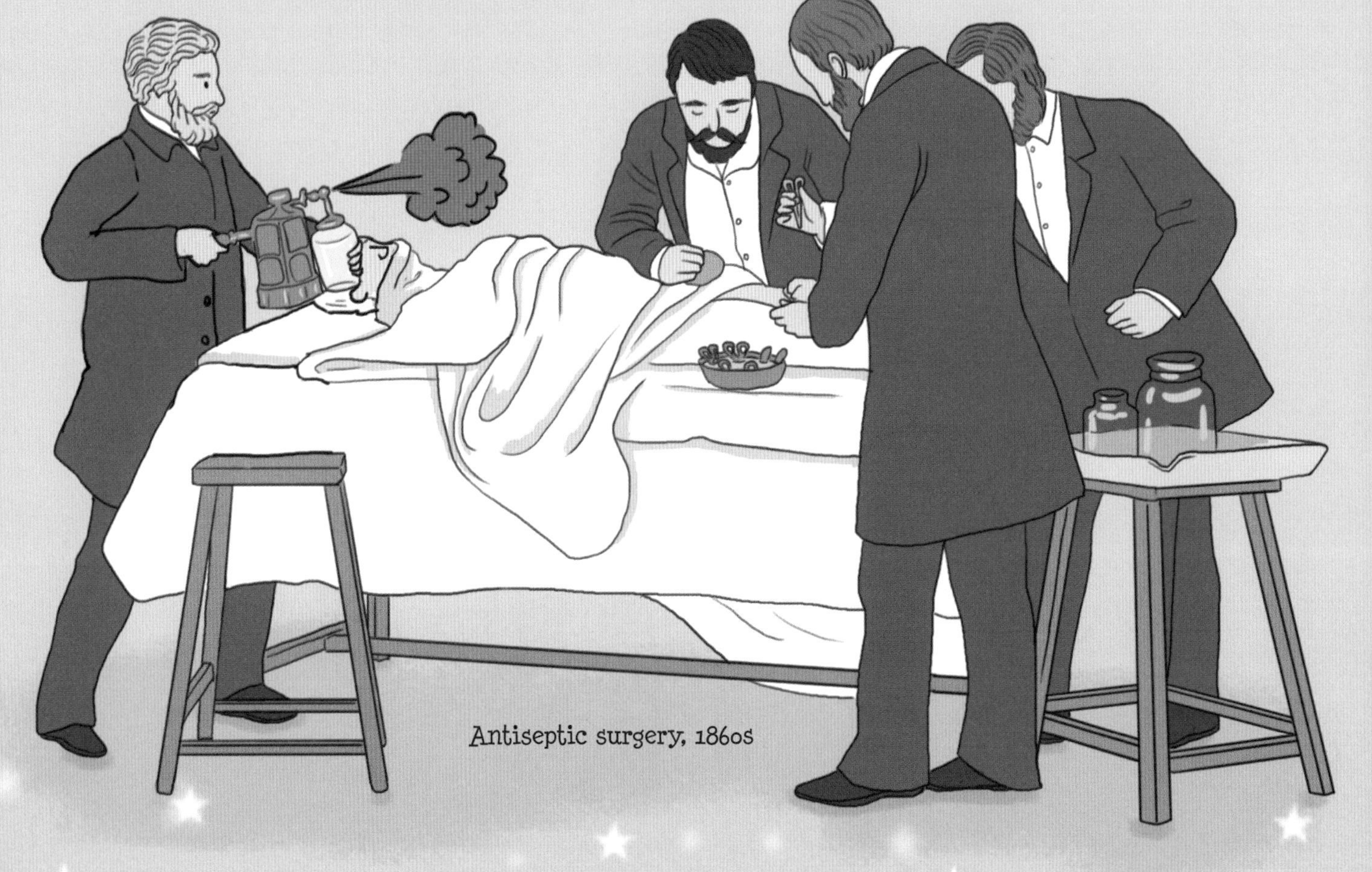
Antiseptic surgery, 1860s

We can't always know who first had the brainwaves behind these inventions. Often, it's just the gradual march of progress. But every once in a while comes a leap of imagination, or a chance discovery, that leads to a new idea. The Popsicle, for instance, was invented by accident when a soda drink was left out on a freezing-cold doorstep!

Bessemer converter, 1856

Making the Modern World takes us through a fascinating history of the inventions that have created our world, from earliest times, through the turmoil of the industrial revolution, to today. Along the way, we'll look at how we grow food, the rise of steam power and electricity, keeping track of time, inventions around the home, medical milestones, smart manufacturing, and the creation of wonder materials like Velcro and Kevlar. And hopefully you'll spot that anyone – including you – can change the way we live today through a clever idea.

Water frame, 1767

Velcro, 1955

AGRICULTURE

Farming today involves high technology, big tractors and complex soil science – but it still relies on the ancient inventions, such as the plough and windmill, with which our ancestors first tamed wild land and made it productive.

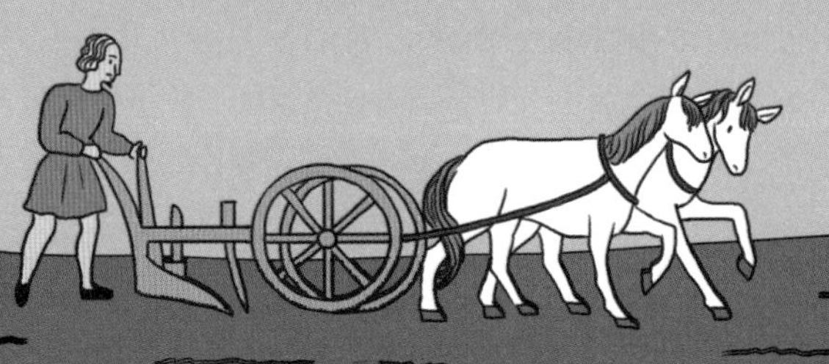

SIMPLE PLOUGH

Once we had domesticated cattle, perhaps 8,000 years ago, the ard followed. This simple plough made from wood, and later metal, scratched single furrows in the soil.

HEAVY PLOUGH

On this ninth-century heavy plough, a coulter makes the first cut in the soil. Behind that is the share, which turns the soil, helped by the mouldboard.

SHADUF

Ancient Egyptians used the shaduf to draw water. On the short end, a weight of clay or stone balanced the bucket end, making it easy for one person to lift.

Hey chum, jump on and I'll give you a lift.
What about us?

WHEELBARROW

Who invented the wheelbarrow? No one's quite sure. Possibly it was the ancient Greeks. But certainly the early Chinese were using it by around 200 AD. (It was the Chinese, too, who first domesticated pigs and chickens.)

LIGHTWEIGHT PLOUGH

This plough of 1730, designed in Rotherham, England, by Joseph Foljambe, was simple to build, and its iron-clad mouldboard made a clean cut.

TRACTOR

Today, you can hitch a multi-bladed plough (or any other farming tool) to the back of a tractor. The tool uses power supplied by the tractor's engine.

WINDMILLS

The first windmills were probably invented in eighth-century Persia for grinding grain or drawing water. They worked 'sideways', turning horizontally inside walls that funnelled the wind through the vanes.

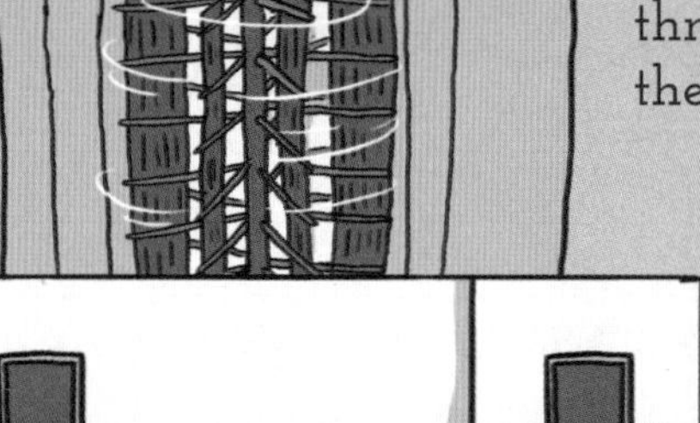

ARCHIMEDES SCREW

The Archimedes screw lifts water by turning a spiral. It's named after the Greek genius Archimedes (287–212 BC), but may have been invented centuries before he was born.

WATERWHEELS

Some 2,000 years ago at Barbegal in France, the Romans built a 'staircase' of 16 connected waterwheels. (The picture shows just the bottom wheel.) They used the mill to grind grain into flour.

FOOD TECHNOLOGY

Your kitchen pantry probably contains cans and jars of preserved food. And you'll have devices to keep food cool (a fridge) or hot (a Thermos flask). But, centuries ago, there were no such conveniences – not even proper cookers. So who invented them all?

FIRST CAN

In 1810, Frenchman Nicolas Appert came up with the first 'can'. He sealed food in a glass jar capped with cork and wax, and boiled it. It would then stay fresh. To prove the method, he once preserved a whole sheep!

Duh, no label – hope it's peas, not dog food.

RING PULL

The ring pull on a drink can was invented in the USA in 1959 by Ermal 'Ernie' Fraze. Within two decades, his company was earning half a billion dollars a year from his clever little tab.

TIN CAN

These are the world's first tin cans, made in London in 1813 by the firm of Donkin, Hall & Gamble. Trouble was, the tin opener wasn't invented till 1855! That's why these men are opening the cans with a hammer and chisel.

THERMOS

Around 1892, Scottish chemist James Dewar invented the vacuum flask, for keeping liquids cool (or hot). Sadly, he didn't patent the idea, but the German company Thermos did... and the rest is history.

Cool. But what is it again?

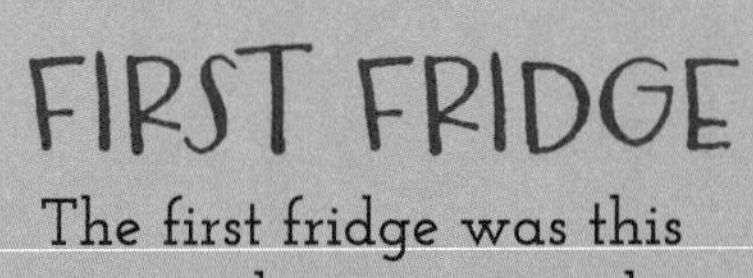

FIRST FRIDGE

What do you think?

The first fridge was this ice machine, invented in America by Oliver Evans and built by Jacob Perkins in 1834. It used a terrifically complicated pumping system to remove heat and cool the water.

FRANKLIN STOVE

In 1741, brilliant American Ben Franklin invented a fireplace with openings that let out more heat without lots of smoke. Modestly, he called it the Franklin stove.

Now will you invent something to scrape it off the floor?

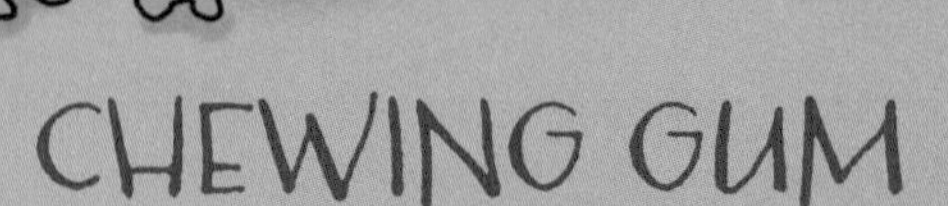

CHEWING GUM

You could cook on a Franklin stove, too. Here, John B. Curtis is heating a pot of sticky resin from spruce trees in 1848, to make America's first chewing gum.

COCA-COLA

In 1886 in Atlanta, Georgia, American chemist John Pemberton began selling a fizzy drink as a 'brain tonic', which he described as 'pure joy'. Its name? Coca-Cola.

IRON AND STEAM

With the invention of the steam engine three centuries ago, and with it the ability to make plenty of good iron and steel, the Western world entered an industrial revolution. Steam, on its own or in partnership with electricity, would power the new factories, steamboats and railway locomotives.

You're selling us down the river!
Oh, just go with the flow.

PAPIN'S STEAM ENGINE

French-born physicist Denis Papin, helped by Gottfried Leibniz, invented one of the first steam engines in 1707, and planned to fit it to a steamboat. But the boatmen on the river Weser in Germany, afraid of the new technology, stopped him carrying out his plan.

PURE IRON

To build the early steam engines, they needed good, pure iron – lots of it. In 1709, Englishman Abraham Darby I worked out how to purify iron by heating it with coke (coal that has been 'cooked'). Newcomen's pumping engine used iron from Darby's furnaces.

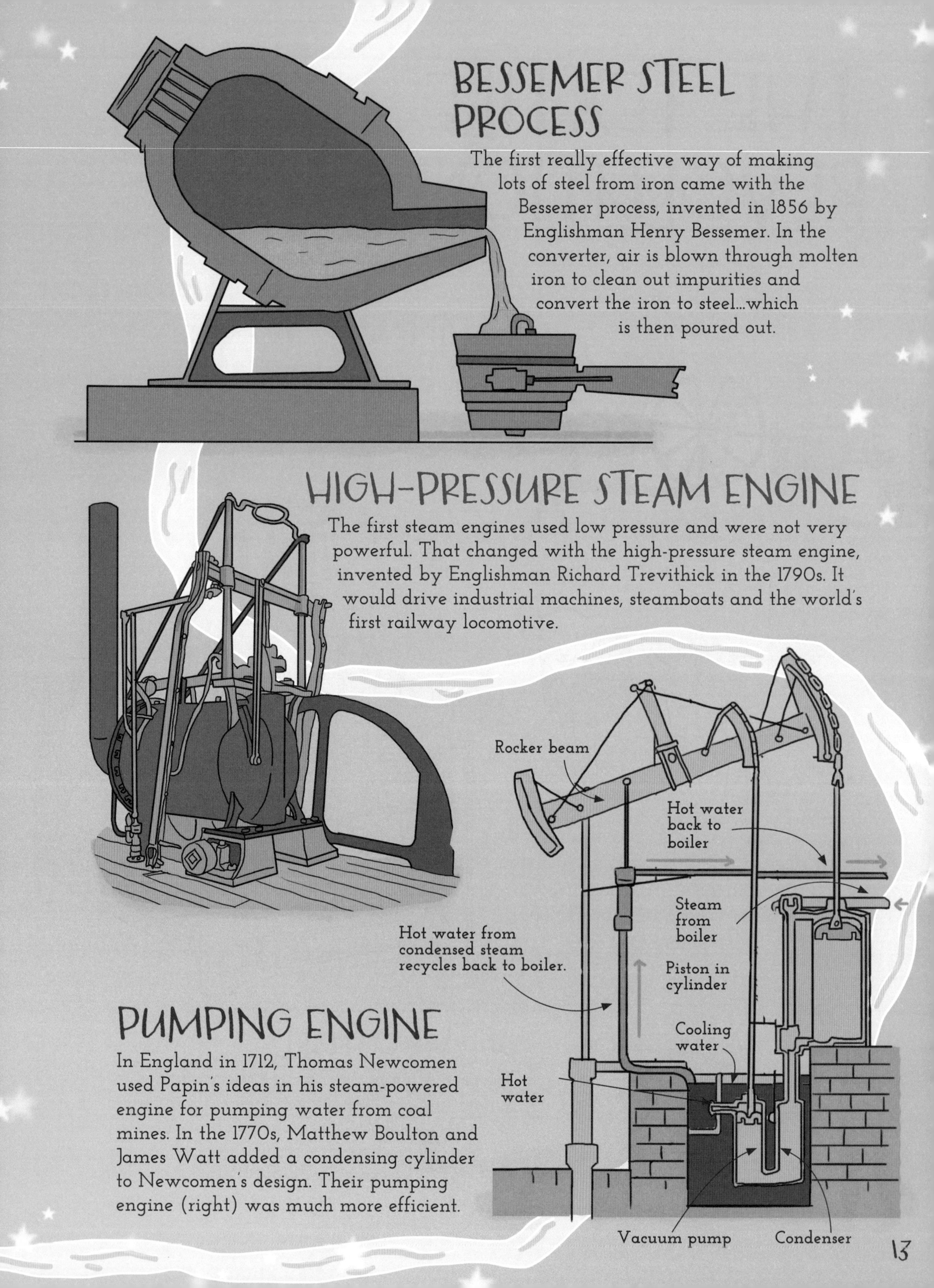

BESSEMER STEEL PROCESS

The first really effective way of making lots of steel from iron came with the Bessemer process, invented in 1856 by Englishman Henry Bessemer. In the converter, air is blown through molten iron to clean out impurities and convert the iron to steel...which is then poured out.

HIGH-PRESSURE STEAM ENGINE

The first steam engines used low pressure and were not very powerful. That changed with the high-pressure steam engine, invented by Englishman Richard Trevithick in the 1790s. It would drive industrial machines, steamboats and the world's first railway locomotive.

PUMPING ENGINE

In England in 1712, Thomas Newcomen used Papin's ideas in his steam-powered engine for pumping water from coal mines. In the 1770s, Matthew Boulton and James Watt added a condensing cylinder to Newcomen's design. Their pumping engine (right) was much more efficient.

THE TEXTILE REVOLUTION

The steam engines of the industrial revolution, along with water power, drove new machines for making goods, such as woven cloth, in greater quantities than ever before. Society changed dramatically as country folk crowded into towns to work in the new factories.

SPINNING WHEEL

In earlier times, making yarn from wool was slow work. The spinning wheel, invented around 1000 AD in India, sped things up, but spinners still found it hard to supply weavers with enough yarn.

AMERICAN WEAVING

Industrial weaving came to the US thanks to Francis Lowell, who had secretly copied British loom designs during a visit in 1810-12. He built textile mills in Massachusetts and fitted them with his new, improved looms.

JETHRO TULL

The revolution in mechanized work reached agriculture, too. The English seed drill, invented in 1701 by Jethro Tull, planted seeds in holes at the right growing depth. (Before, seeds had just been thrown across the soil.)

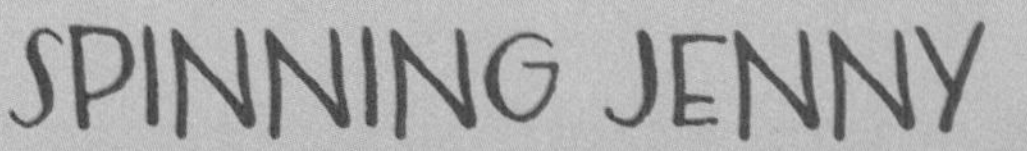

SPINNING JENNY

In 1764, Englishman James Hargreaves invented the spinning jenny, a machine that could spin several spools of yarn at once. One machine did the job of eight, ten or even 100 spinners.

The machine's name, 'jenny', probably came from 'engine'.

Why do they keep calling me Jenny? I'm Maggie.

WATER FRAME

Richard Arkwright's water frame (1767) was a yarn spinner powered by water flow. In his Nottingham factory, belts, pulleys and shafts connected a waterwheel to each floor, driving the frames.

ELECTRICITY

Electricity has a long history. Early peoples knew about electric fish, and lightning, and static electricity (the crackly tingle you make by stroking your cat's fur, for instance). But storing electricity, and using it as a power source, dates from 350 years ago.

LEYDEN JAR

This is a Leyden jar. It's a very crude battery or capacitor, capable of storing electricity, invented around 1745 by Ewald von Kleist in Germany and Pieter van Musschenbroek in Leiden, Holland.

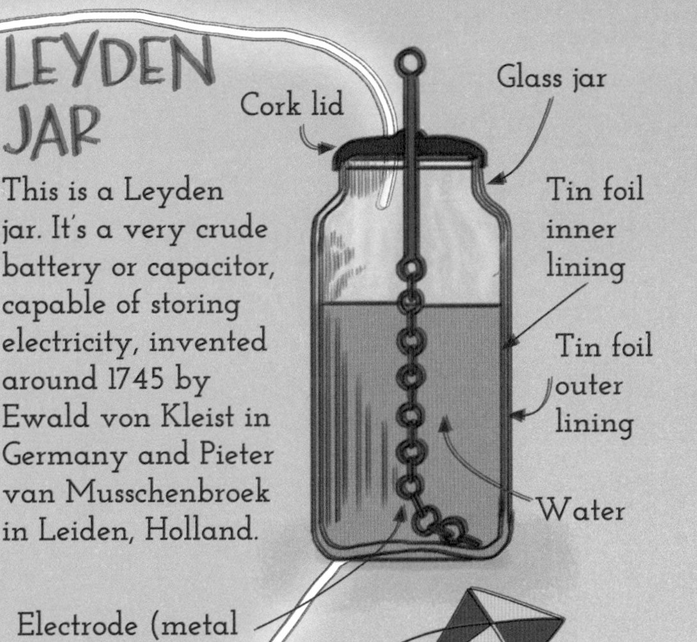

TESLA COIL

Serbian Nikola Tesla (1856-1943) pioneered radio, neon lights, alternating current, X-rays, wi-fi and even a 'death ray' (which luckily he didn't build). His Tesla coil is a kind of transformer.

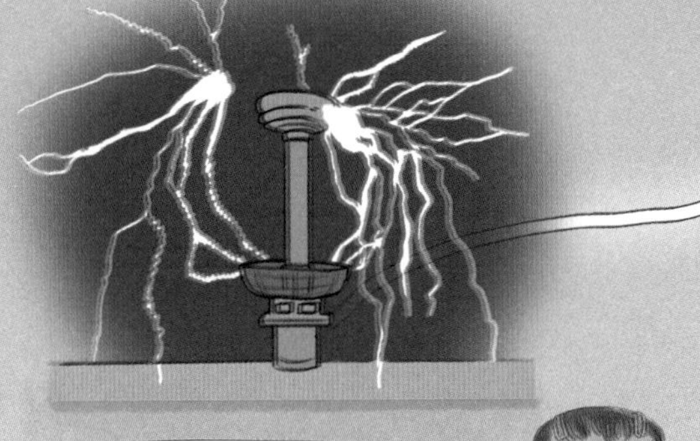

CATCHING LIGHTNING

It's said that in 1752 Ben Franklin, or his son, flew a kite to 'catch' lightning and store it in a Leyden jar, but he probably never did. He realized, though, that lightning equals electricity. And he invented the lightning rod, and the word 'battery'.

You can wire me to the mains!

VOLTAIC PILE

Italian scientist Alessandro Volta invented his 'voltaic pile', a battery, in the 1790s. His countryman Luigi Galvani had earlier found that if you sent electricity through a dead frog's legs, they twitched – showing that nerve signals are electric 'messages'.

ELECTRIFYING INVENTOR

Frenchman Gustave Trouvé (1839-1902) designed this glowing jewellery. The list of his electrified inventions includes a canoe, outboard motor, dentist's drill (ouch), metal detector, airship, razor, sewing machine, microphone, telegraph, rifle... and more.

My bling may be 'light', but it feels sooo heavy!

LIGHTNING ROD

The 19th century saw a craze for all things electrical. This man has a Franklin rod attached to his umbrella to ward against lightning strikes.

HYDROELECTRIC POWER

The first house ever to run on hydroelectric power was Cragside in England. Its dynamos provided lighting, and powered the nearby farm buildings.

ELECTRIC MOTOR

In 1869, Zénobe Gramme invented an electric generator or dynamo. It wasn't very good – but he later found (by accident) that his machine, if connected to an electric source, would spin by itself. He had invented the modern electric motor.

FARADAY DISK

English scientist Michael Faraday made the crucial connection between electricity and magnetism. He invented an early electric generator, the Faraday disk, in 1831. It works by spinning a metal flywheel in a magnetic field.

Oops! But then again, hurrah!

MEASURING TIME

We've come up with all sorts of inventions to measure the passage of time. The ancients saw the sun rise and set each day, and very sensibly used it to measure the hours. Of course, sundials don't work in the dark, so the clockwork mechanism was the next leap forward.

3D SUNDIAL

This 3D sundial, invented in 250 BC by Greek astronomer Aristarchus, is called a scaphe. The post casts a shadow inside the hemisphere, showing the time. (Aristarchus, by the way, used it to calculate Earth's size, and its distance from the Moon.)

WATER CLOCK

The ancient Egyptians relied on the water clock, or clepsydra: water dripped from one container into another to measure the hours. (They used it to time people's speeches during debates. No more drip – zip your lip!)

Oh look, it's time for tea!
TICK TICK TOCK TICK
Humph! Tea break is over.

SUPER TIMER

A water clock designed in the third century BC by Ctesibius of Alexandria is said to have been the most accurate timekeeper of the next 1,800 years.

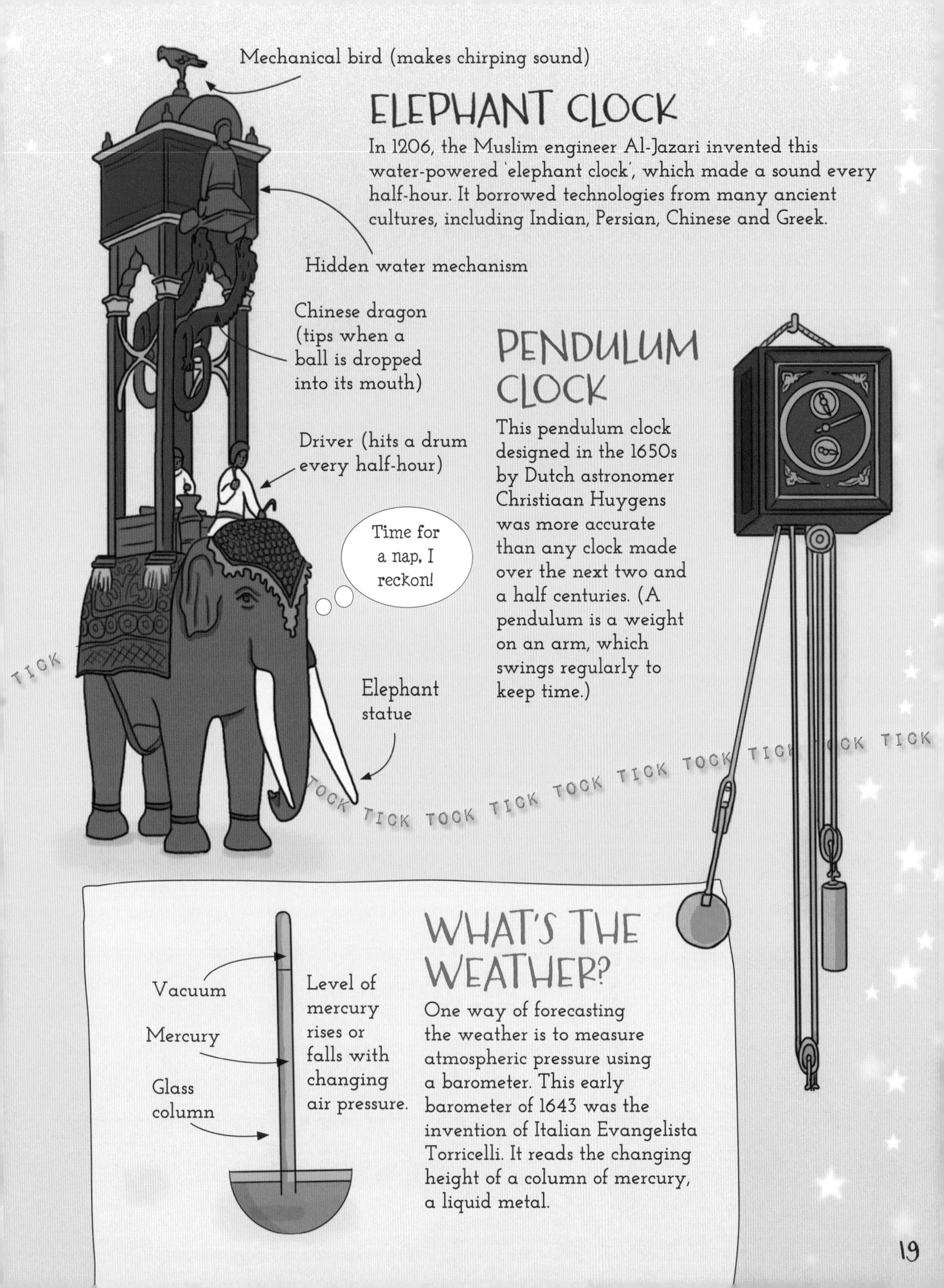

ELEPHANT CLOCK

In 1206, the Muslim engineer Al-Jazari invented this water-powered 'elephant clock', which made a sound every half-hour. It borrowed technologies from many ancient cultures, including Indian, Persian, Chinese and Greek.

PENDULUM CLOCK

This pendulum clock designed in the 1650s by Dutch astronomer Christiaan Huygens was more accurate than any clock made over the next two and a half centuries. (A pendulum is a weight on an arm, which swings regularly to keep time.)

WHAT'S THE WEATHER?

One way of forecasting the weather is to measure atmospheric pressure using a barometer. This early barometer of 1643 was the invention of Italian Evangelista Torricelli. It reads the changing height of a column of mercury, a liquid metal.

AT HOME

Take a look around your home and school. Some of the basic materials – glass, concrete, brick – date back to ancient civilizations, and it's hard to say exactly who invented them. But all those nifty devices, such as locks, doorbells, elevators and toilets, come from someone's brainwave.

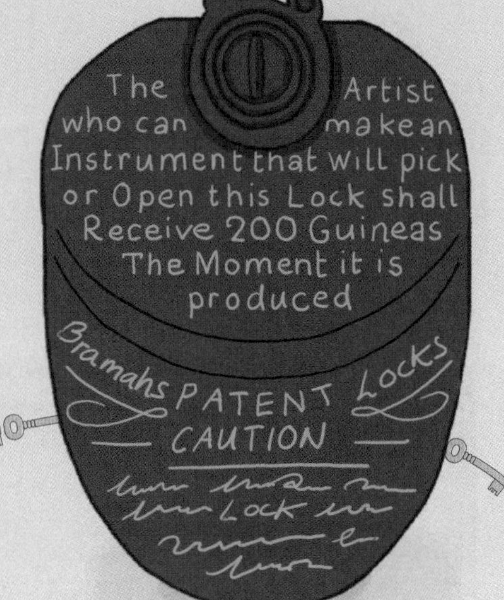

The Artist who can make an Instrument that will pick or Open this Lock shall Receive 200 Guineas The Moment it is produced
Bramahs PATENT Locks
CAUTION

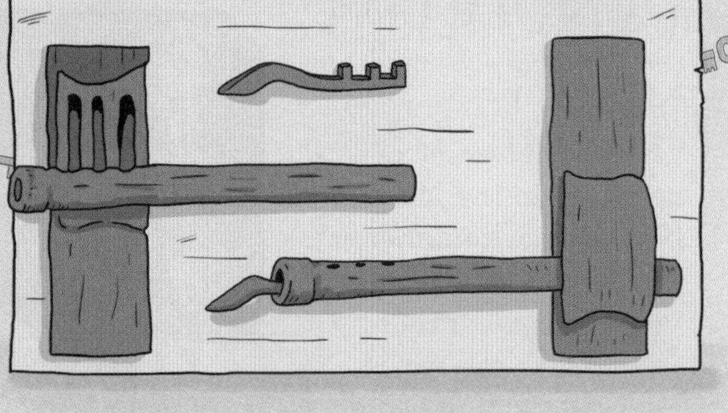

PIN LOCK

Over 2,000 years ago, the Mesopotamians used key-operated door locks, made from timber. This is an ancient Egyptian pin lock.

CHALLENGE LOCK

High-security locks date from 1784, when British engineer Joseph Bramah designed his 'Challenge' lock. He promised a big cash prize to the first person to pick it (unlock it without a key). It remained unpicked for 67 years! By then, of course, old Joe had died.

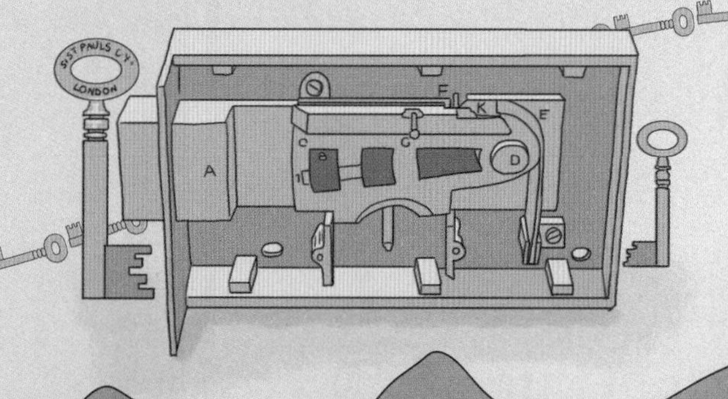

CHUBB LOCK

English locksmith Jeremiah Chubb invented this 'detector lock' in 1818. If you used the wrong key in it, the lock jammed and could only be opened with a special extra key (shown on the left).

ELECTRIC DOORBELL

In 1831, American scientist Joseph Henry invented a doorbell that rang inside a building via an electric wire. Yep: we have him to thank for all those kids who ring the bell and run away. (The electric relay in Joe's bell was later used by Samuel Morse in the Morse code tapper.)

I could get tired of this.

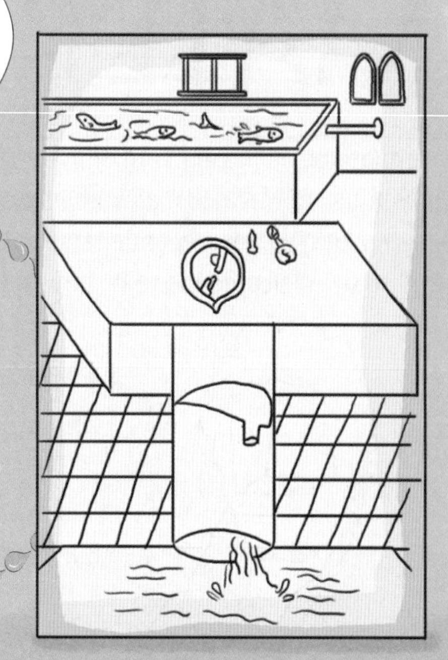

TOILETS

If there's one home invention we use every day, it's the loo. It's an old one, too. The Romans had toilets with drains to flush away the poo, and they wiped their bums with a sponge on a stick.

FLUSHING TOILETS

In 1596, English writer Sir John Harington drew this design for the 'Ajax', the world's first true flushing loo. (In his day it was called 'the jakes'.) He probably didn't really keep fish in the water tank, though!

BUILT TO LAST

Locksmith Joseph Bramah also designed flushing toilets. Some of those he built in his London workshop are still working today.

SAFETY LIFT

Early elevators were dangerous: if their lifting cable broke, they could hurtle to the ground, killing the passengers. So, American Elisha Otis invented a safety lift with an emergency brake, and demonstrated it at New York's World Fair in 1854. The cable was cut, and Elisha fell just a few inches in his lift before the brake stopped it. Success!

MATERIALS

Perhaps you've never heard of materials science, but it's big business. Inventors are constantly at work improving the stuff things are made of – making it cheaper, better at its job, better for the environment and so on. Welcome to the history and science of 'stuff'!

MACKINTOSH

The 'Mackintosh' raincoat was invented by Scotsman Charles Macintosh in the 1820s. He made the waterproof fabric by sandwiching rubber between two layers of cloth.

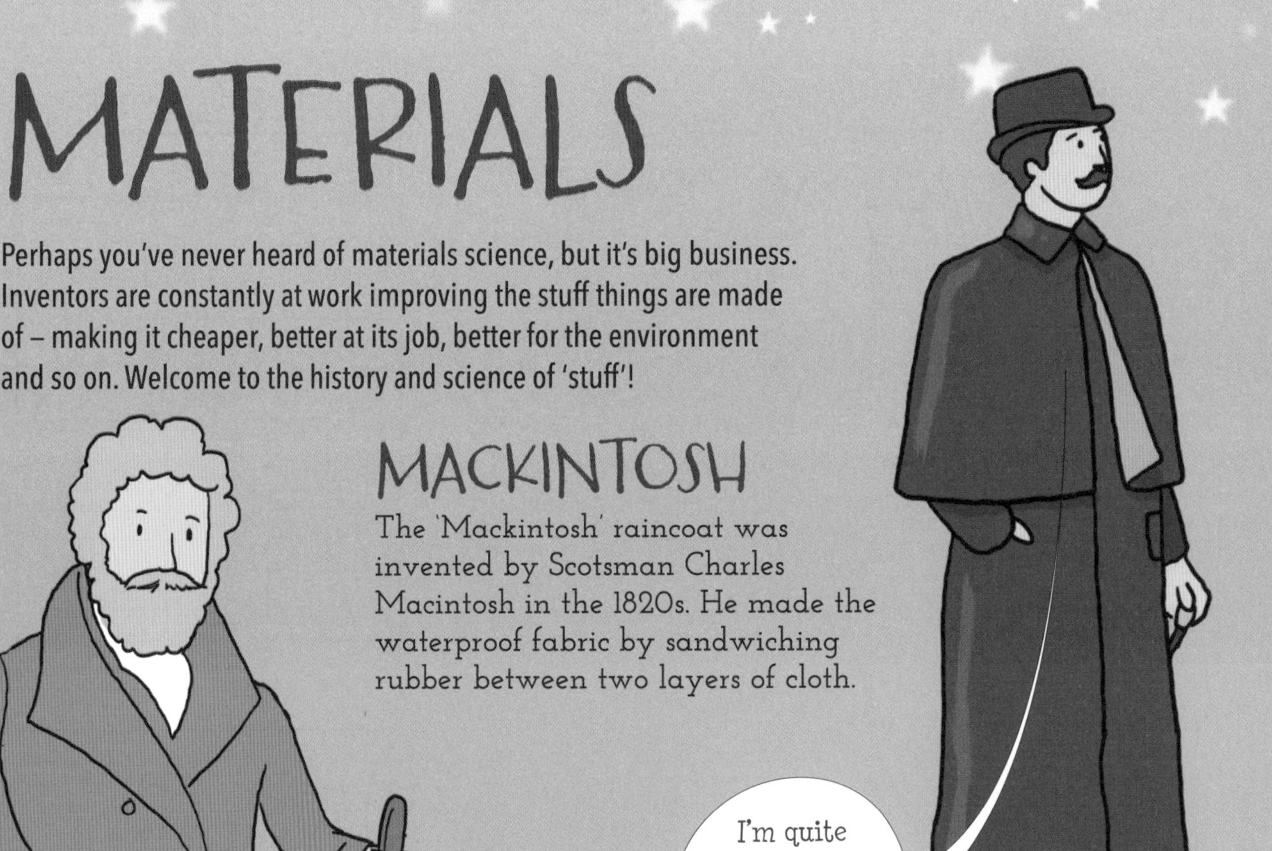

I'm quite dry (so no drippy jokes please!).

VELCRO

Velcro was invented by a dog. Nearly. George de Mestral was walking in Switzerland in 1941, and noticed how the hooked spines of plant burrs stuck to his dog's fur. That gave him the idea for a clingy strip-fastener with one side of 'fur' and another of tiny hooks. Velcro was launched in 1955.

Yup, I invented this!

DUPONT INVENTIONS

Workers at the American company DuPont have a knack of inventing new materials: here's a roundup of some of their most famous.

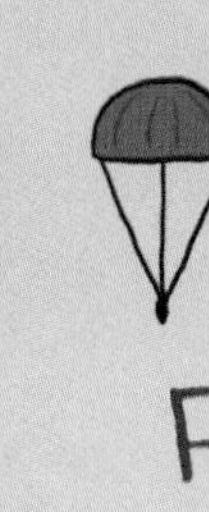

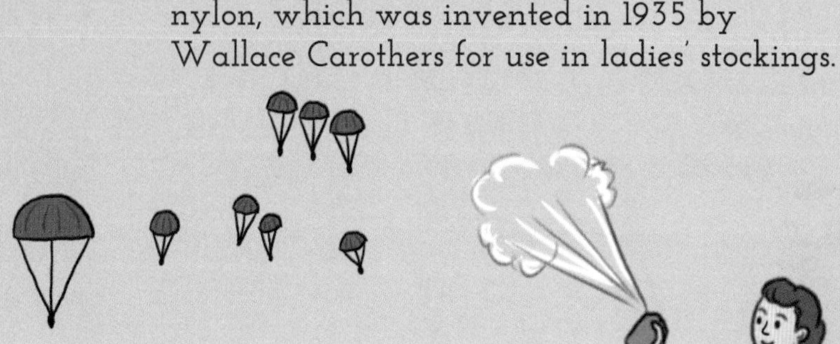

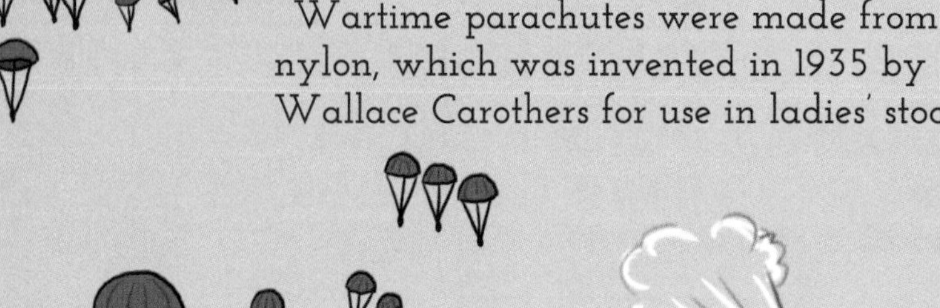

NYLON

Wartime parachutes were made from nylon, which was invented in 1935 by Wallace Carothers for use in ladies' stockings.

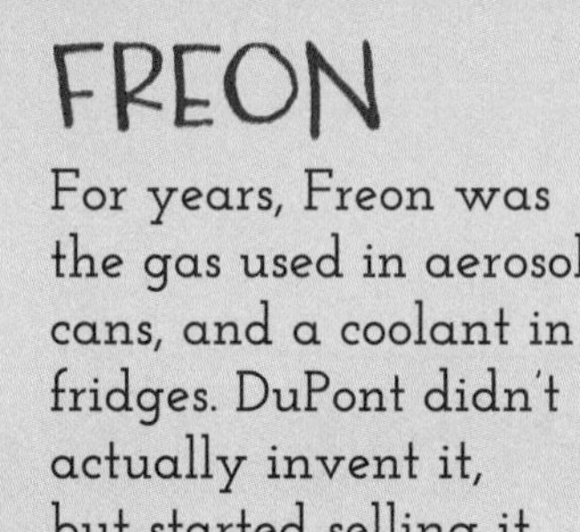

FREON

For years, Freon was the gas used in aerosol cans, and a coolant in fridges. DuPont didn't actually invent it, but started selling it in 1930. Today, Freon isn't used much, as it contributes to climate change.

TEFLON

Teflon is the non-stick pan coating that helps your parents flip omelettes. Easier to say than polytetrafluoroethylene, it was discovered in 1938 by Roy Plunkett.

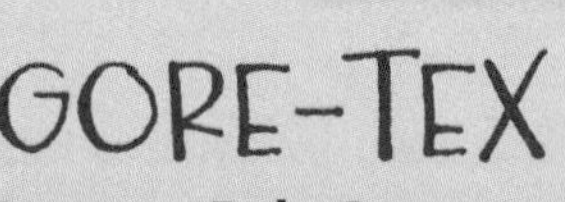

GORE-TEX

In 1969, Bob Gore stretched a piece of Teflon to 10 times its normal length... and found that it turned into a new fabric that was waterproof, yet breathable. We know it as Gore-Tex, and it's used in raincoats and walking boots.

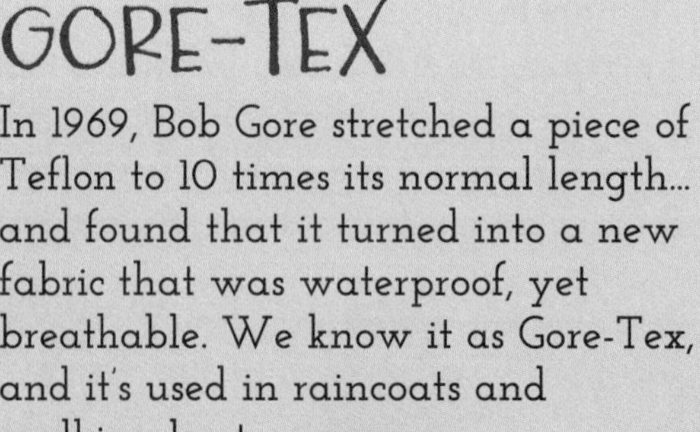

KEVLAR

This police dog wears bullet-proof armour made from Kevlar, a super-tough material invented by chemist Stephanie Kwolek in 1965.

MEDICINE

No book, however big, could be big enough to cover all the fascinating stories of medical discovery and invention. So here are just a few medical milestones – from studying the human body, to pain relief, the prevention of disease and simple good hygiene.

STUDYING ANATOMY

Dutchman Andries van Wesel (1514-64), better known as Vesalius, was one of the first scientists to dissect (cut open) dead bodies to see how they worked. He published books full of anatomical drawings.

Can I move now? I've been sitting here for 12 years!

FIRST VACCINE

In 1796, English scientist Edward Jenner injected a boy with cowpox, collected in pus from a milkmaid's blisters. It protected the boy against smallpox, which was a killer disease. Thanks to Jenner's pioneering vaccination work, smallpox is no more.

STETHOSCOPE

French doctor René Laënnec invented the stethoscope in 1816, when trying to listen to a woman's heart. He rolled up some paper into a tube, put his ear to one end, and heard it clearly. Next, he made a wooden tube. Today's stethoscopes look quite different, but they work the same way.

FALSE TEETH

False teeth were invented centuries ago, possibly in Italy, but some of the most famous were those of US president George Washington (1732-99). They included teeth from hippos, cows, other people, and maybe also elephants!

LOUIS PASTEUR

The study of germs, disease and vaccines by French chemist Louis Pasteur has saved countless lives. In 1862, he invented (and gave his name to) pasteurization: heating milk and beer to kill germs. Here, Pasteur looks on as his colleague, Emile Roux, vaccinates a boy against rabies.

Say aaarggh!

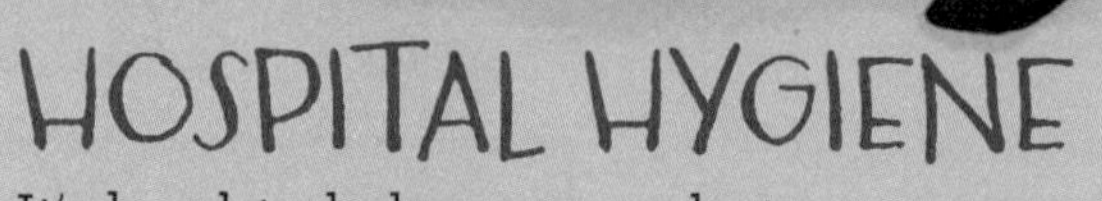

HOSPITAL HYGIENE

It's hard to believe now, but surgeons used not to wash their hands before performing operations. In 1847, Hungarian doctor Ignaz Semmelweis believed such uncleanliness was somehow spreading germs and killing patients, and he recommended hand-washing. In doing so, he more or less 'invented' hospital hygiene.

The soap's in here somewhere...

ANTISEPTIC SURGERY

From 1867, inspired by Pasteur and Semmelweis, British surgeon Joseph Lister 'cleaned up' operations by spraying his patients with carbolic acid, an 'antiseptic' or disinfectant. It killed germs and prevented infection setting in.

FACTORIES

It's one thing to invent better products; it's quite another to invent better ways of making them. The mass-production factory has a long history, but these days we're realizing that big factories can cause waste and pollution. So we're beginning to make smarter.

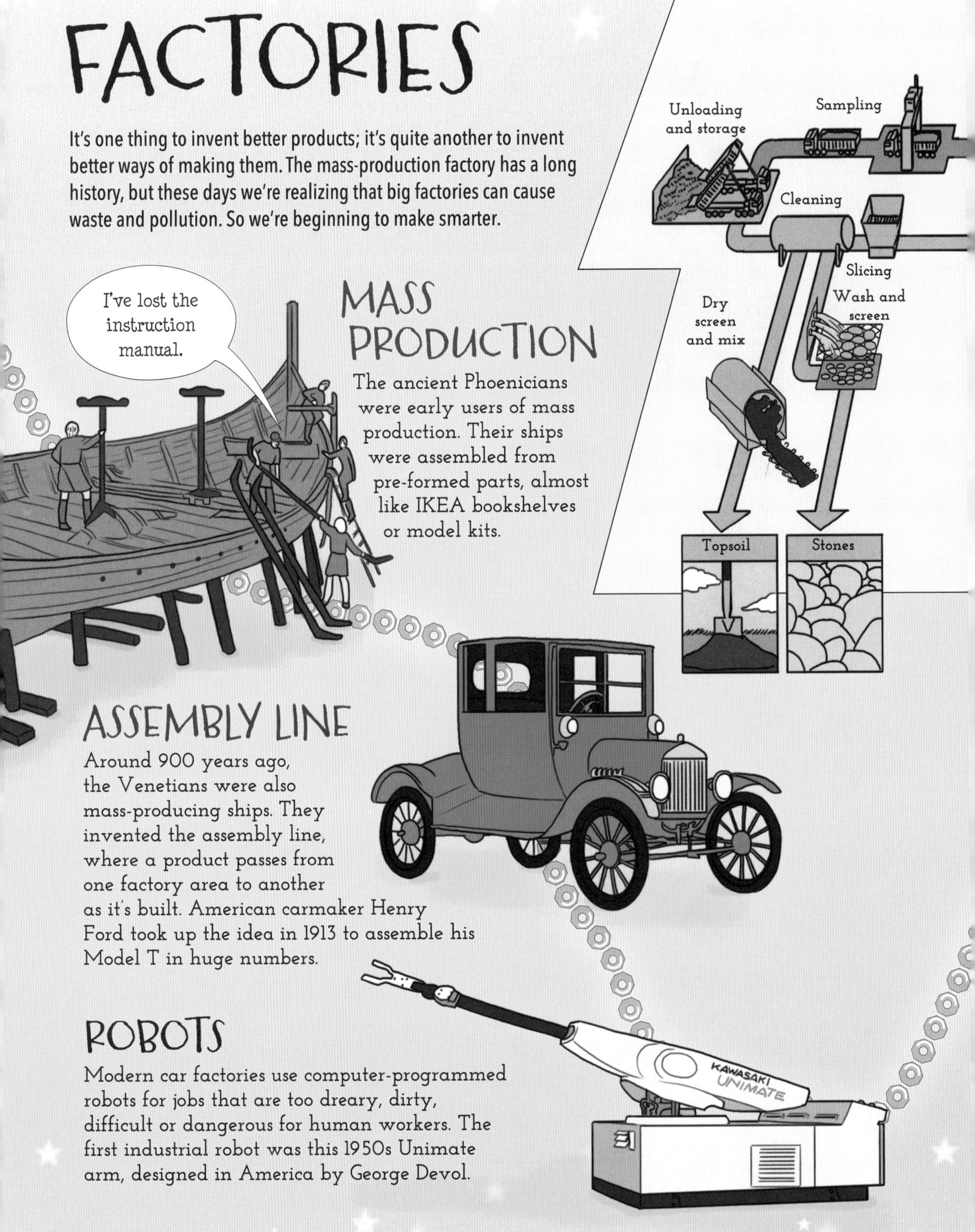

MASS PRODUCTION

The ancient Phoenicians were early users of mass production. Their ships were assembled from pre-formed parts, almost like IKEA bookshelves or model kits.

ASSEMBLY LINE

Around 900 years ago, the Venetians were also mass-producing ships. They invented the assembly line, where a product passes from one factory area to another as it's built. American carmaker Henry Ford took up the idea in 1913 to assemble his Model T in huge numbers.

ROBOTS

Modern car factories use computer-programmed robots for jobs that are too dreary, dirty, difficult or dangerous for human workers. The first industrial robot was this 1950s Unimate arm, designed in America by George Devol.

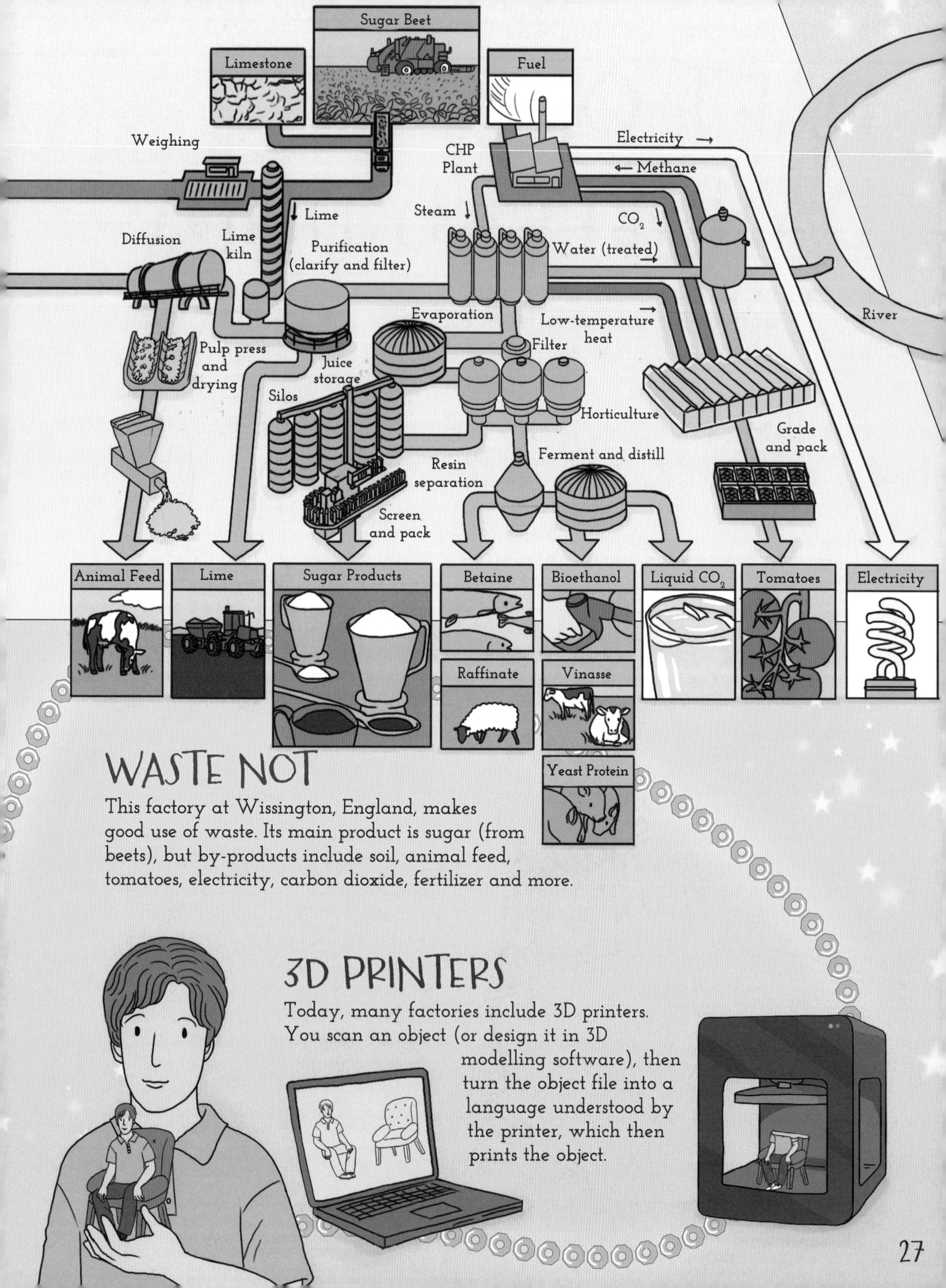

WASTE NOT

This factory at Wissington, England, makes good use of waste. Its main product is sugar (from beets), but by-products include soil, animal feed, tomatoes, electricity, carbon dioxide, fertilizer and more.

3D PRINTERS

Today, many factories include 3D printers. You scan an object (or design it in 3D modelling software), then turn the object file into a language understood by the printer, which then prints the object.

DAFT OR DANGEROUS DUDS

Anybody can come up with new ideas for gadgets and toys, whether they're a famous inventor or someone like you. But that doesn't mean every new idea is a good one – as these examples show!

TALKING DOLLS

Thomas Edison, pioneer of light bulbs, record players and movie cameras, had some bad ideas, too. He designed these big dolls for kids in 1890. Each doll contained a gramophone that 'sang' nursery rhymes. But they were expensive, and made hideous noises that terrified the customers, who sent them back to the shop. 'Little monsters', Ed later called them.

Whaddya mean, this is a wobbly idea?

JELLY-MOULD HOUSES

Back in 1899, poor people needed housing. Ed thought he'd help by casting cheap concrete houses from moulds, almost like giant jellies. Only they weren't cheap: they cost 40 times more than normal houses! Today, just one Edison concrete home survives, in Montclair, New Jersey.

Not what I'd call a walk in the park...

HOTTER TROTTER

This 1868 'steam man' was designed by Zadoc Dederick to pull a carriage. It was dressed as a man so as not to frighten horses on the street. It was worse than useless, but looked fabulous. The top hat cleverly hides the steam exhaust pipe, by the way.

QUAKE-MAKER

Nikola Tesla, electrical genius, also had his off-days. He once invented an earthquake machine, which could (in theory) bring buildings down. He considered it so dangerous, he smashed it with a hammer and never spoke of it again.

BABY ARMOUR

In Britain, the threat of an enemy gas attack in the lead-up to World War II led a Mr Mills to invent this Air Raid Precautions gas-proof baby buggy. There was a window on top so mum could check baby was still breathing.

ACCIDENTAL INVENTIONS

The popsicle was invented by mistake. Here are a few more happy accidents.

PENICILLIN

Scottish biologist Alexander Fleming discovered the disease-killing drug penicillin in 1928 after his nose dripped on a specimen dish in his lab, and it grew mouldy.

Bless you, guv'.

SLINKY TOY

American Navy engineer Richard James invented Slinky in 1943. It was going to be a suspension spring for warships, but he found it righted itself when tipped over.

Apparently he's invented a super new glue.

SUPER GLUE

In World War II, American chemist Harry Coover was trying to make gunsights and aircraft canopies out of cyano-acrylate, a clear resin. No such luck – but it was really sticky! Today we call it contact adhesive, or 'super glue'.

POTATO CRISPS

In 1853 in New York, a diner sent his French fries back, complaining they were too thick. So the chef, George Crum, got grumpy and sliced them ultra-thin, baked them hard, and coated them with salt, intending to make them inedible. But the diner loved them! George had invented crisps.

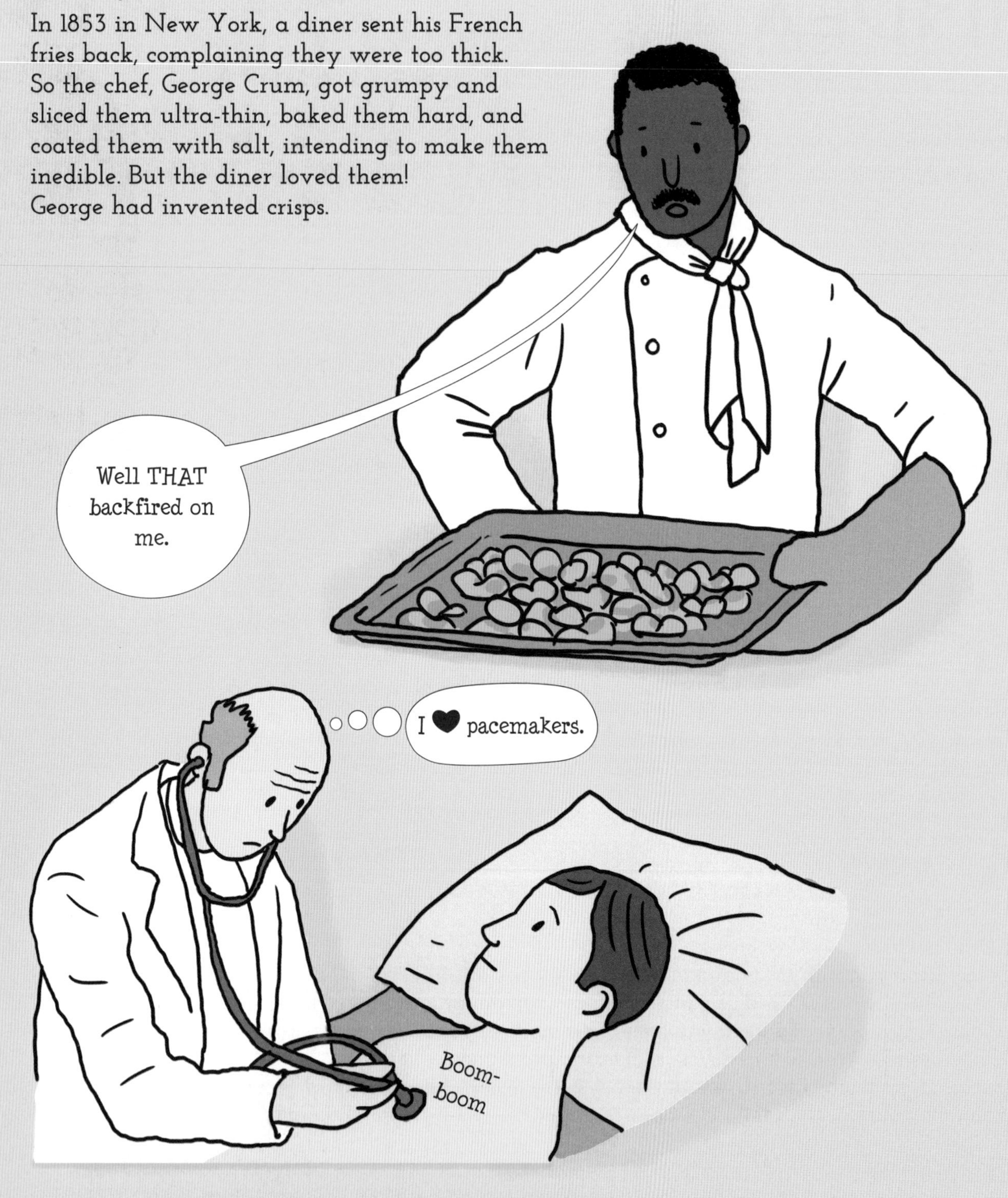

IMPLANTABLE PACEMAKER

A pacemaker is an electronic implant that helps unsteady hearts keep a regular beat. American electrical engineer Wilson Greatbatch invented it by mistake in 1956 – he used the wrong component while building a recording device. First used in 1960, his pacemaker has since prolonged countless lives.

INDEX

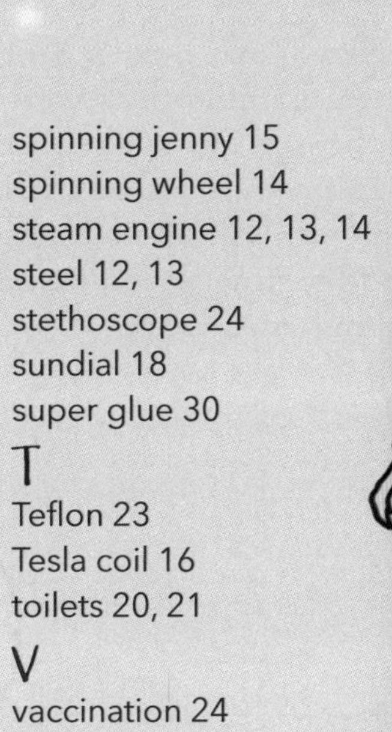

The Author

British-born Matt Turner graduated from Loughborough College of Art in the 1980s, since which he has worked as a picture researcher, editor and writer. He has authored books on diverse topics including natural history, earth sciences and railways, as well as hundreds of articles for encyclopedias and partworks, covering everything from elephants to abstract art. He and his family currently live near Auckland, New Zealand, where he volunteers for the local Coastguard unit and dabbles in art and craft.

The Illustrator

Sarah Conner lives in the lovely English countryside, in a cute cottage with her dogs and a cat. She spends her days sketching and doodling the world around her. She has always been inspired by nature and it influences much of her work. Sarah formerly used pens and paint for her illustration, but in recent years has transferred her styles to the computer as it better suits today's industry. However, she still likes to get her watercolours out from time to time, and paint the flowers in her garden!

INCREDIBLE INVENTIONS

MAKING THE MODERN WORLD

Written by Matt Turner
Illustrated by Sarah Conner

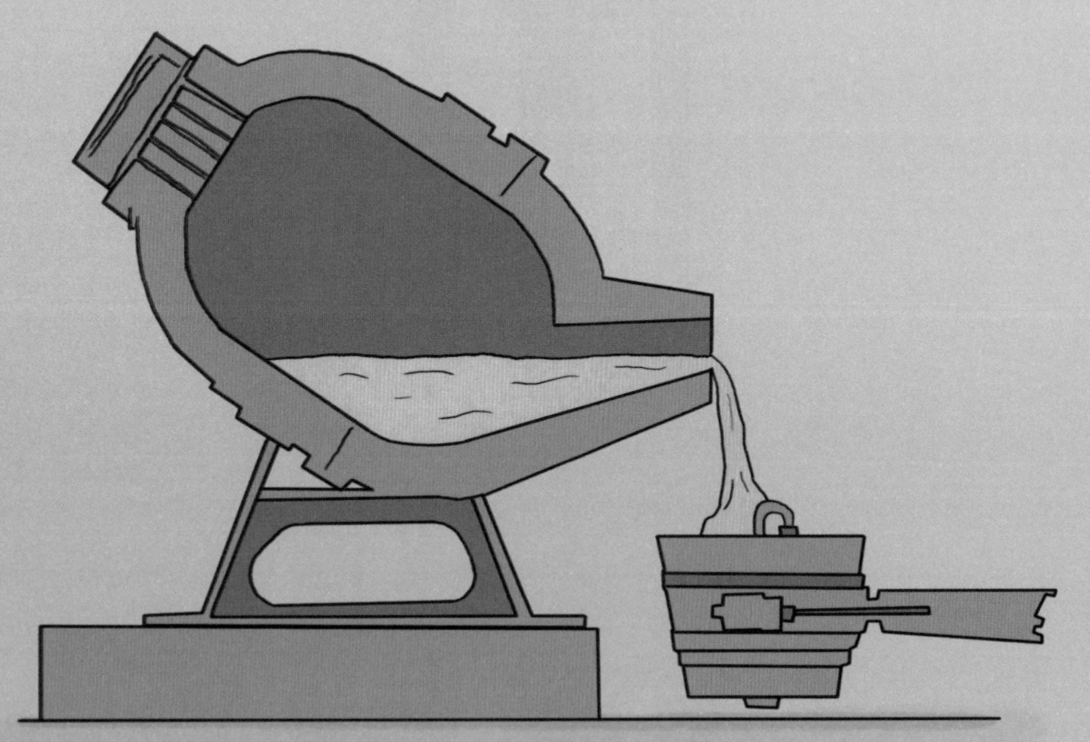

外语教学与研究出版社
FOREIGN LANGUAGE TEACHING AND RESEARCH PRESS
北京　BEIJING

番茄博士双语百科

奇思妙想发明史

人类沟通简史

马特·特纳（英）著
萨拉·康纳（英）绘
刘姣 译

只剩1,285
页就印刷
完毕了……

外语教学与研究出版社
FOREIGN LANGUAGE TEACHING AND RESEARCH PRESS
北京 BEIJING

we're
ON A ROLL!

1,000,000,000

最早的印刷技术（借此技术你才能读到这本书）是中国人发明的，但是人们首先得发明文字符号。去看看文字诞生的故事吧，书里还会向你介绍人类为了沟通而使用的多种方法和机器，比如信号、密码、电话、无线电、电视和计算机。在这本书里，你会发现：

- 谁是电话的真正发明者
- 什么密码用了267年才被人们破解
- 电视的发明者用废品造了早期的电视（他只用得起废品）
- 计算机程序这一理念源于布料编织

还有更多的发现等着你。不过也有一些怪诞的发明（比如狗吠翻译机），以及错得离谱的预测——有人说1996年是因特网的末日，有意思……（提出这个预测的人真把自己说的话给吃了。）

1861年，德国教师约翰·赖斯发明了一部电话。想知道图中这匹马要干什么，请翻到第18页。

目录

导语：从原始洞穴图画到因特网

人类沟通史始于符号系统和文字的发明，中间经历了印刷机、电报机和电话机的诞生，而今进入了广泛使用计算机、卫星和智能手机的数字时代。

苹果手机，2007年

我们最早的祖先在洞穴壁上留下了令人惊艳的图画，也许夜晚他们围着篝火交谈时，也使用了像这些画一样美丽的语言。对此我们无从得知，因为没有关于他们语言的记录——那时还没有可书写的符号系统。甚至到了现代，有些文明仍不使用文字。比如19世纪20年代欧洲人来到新西兰后，当地的毛利人才开始使用书面文字。然而，他们传统的文身和雕刻本身也是一种语言。

纸，公元105年

今天，人际沟通比以往任何时候都要容易和快捷，而且形式多样。我们每天都可能用到电话、电子邮件、短信、社交网站脸书、微博客推特、网络电话讯佳普和照片及视频分享应用Instagram等。有时候，我们即使跟家人同在一个房间，也会给他们发短信！绕了一大圈，我们就这样以一种奇特的

方式又回到了原始洞穴时代，跟国内甚至国外的朋友们（通过电子通信技术）直接通话，而不是给他们写信。

我们一起来看看这些巨变是如何发生的吧。有些变化是慢慢发生的，就像古老的图片文字那样，历经数个世纪。但有些变化发生得很快，比如因特网诞生了还不到50年，而脸书只能算个少年。

传真机，1840年

来了解一下是谁发明了纸笔、印刷术、电子邮件和Instagram吧。还有一些守旧派认为收音机、电视机、电影院和计算机等不会有前途，他们提出的令人捧腹的预测读起来会很有意思。

（收音机里的）真空管，1904年

录像机，20世纪50年代

语言的诞生

自有人类起就出现了语言。但最早的文字来源于艺术：如果你想说“太阳”或“熊”，就画一幅太阳或者熊的图。随着时间的流逝，这些图画慢慢变成被编码的文字系统，也就是符号系统。现在全世界已有6,000多种语言。但是除了诞生时间最晚的语言，我们无法确定其他语言是谁发明的。

最早的文字

大约5,200年前，中东地区的人们开始使用文字，主要用于记录税收和交易。下图这个苏美尔抄写员正用一根芦苇在湿黏土（会慢慢变干）上刻楔形。苏美尔人的这种文字被称为楔形文字。

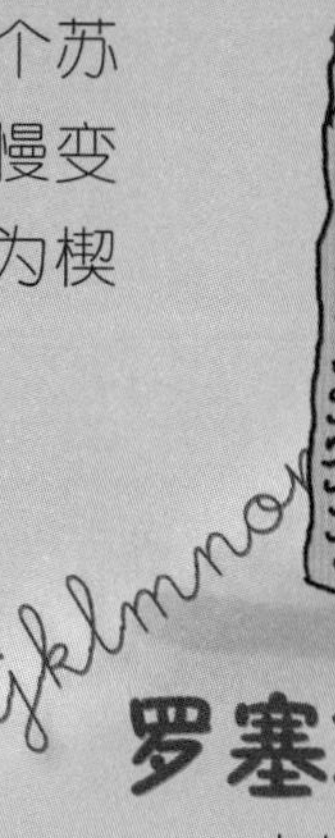

罗塞塔石碑

古埃及祭司使用的图画文字被称为象形文字（“神圣的雕刻”）。在很长的时间里，我们并不知道象形文字的含义。直到1799年，一个法国士兵在埃及发现了一块古老的石碑，上面用三种语言记录了同样的信息。我们终于能弄清楚那些祭司写的是什么了。

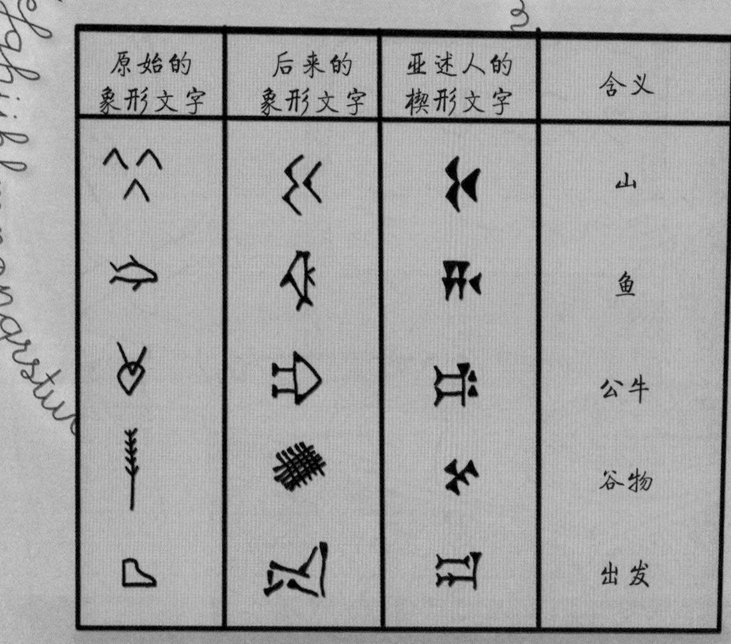

原始的象形文字	后来的象形文字	亚述人的楔形文字	含义
			山
			鱼
			公牛
			谷物
			出发

象形文字

最早开始写作的人使用的是象形文字，即用图形描绘事物形状的一种文字。因此，“水”的象形文字常为一条波浪线，“公牛”的象形文字则是一个长了角的头。后来，人们开始使用更简单的形状，画起来比象形文字更快。左边这个表展示了古代中东地区一些早期的文字系统。

皮特曼速记法

时间退回到19世纪30年代的英格兰，当时一位名叫艾萨克·皮特曼的教师认为人们写字太费时间了。他说："节约时间就是延长生命。"千真万确！艾萨克发明了一种用简单的线条和不规则的曲线组成的速记语言。每种形状代表一种发音或一个简单的词（比如"那个"或"你"）。今天，皮特曼速记法仍然很流行，还在为人们节省时间。

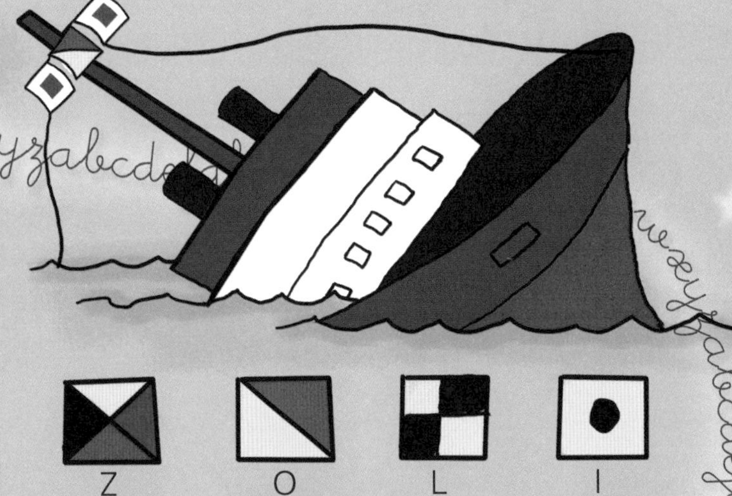

信号旗

无线电发明以前，水手像晾衣服那样把旗帜挂在船的桅杆上来发送信号。弗雷德·马里亚特船长在1817年发明了一种基础的信号旗。40年后，这种信号旗发展成了由26种旗帜组成的国际信号旗系统。你只用一面旗就可以发一个信号，如A字旗表示"水下有潜水员"。或者也可以将许多面不同的旗串联起来组成词语。

Z O L I

K E S H

世界语

世界语是波兰医生路德维希·柴门霍夫于1887年发明的一种语言。他想尽量简化这种语言，使之成为人人都能使用的国际语言。现在，全世界有超过200万人会说世界语。

书写工具的发明

文字的载体和文字同等重要。你总不能把石碑带到商店里去吧！早期人类用植物纤维制作书写材料，也用动物皮革制作羊皮纸。纸是很久以后才发明的，铅笔和钢笔的历史就更短了。

甲骨文

迄今发现的最早的文字是在中国贾湖发掘出的一组文字符号，大约在8,500年前被刻在龟壳和骨头上。没人能确定这些文字符号的含义，也许是魔咒吧。

纸

古埃及人把纸莎草这种沼泽植物撕碎捣烂，制成平整的、用于书写的纸片，我们才有了“纸”这个词。大约在公元105年，中国人蔡伦用破布、树皮和旧渔网造出了纸。

墨

中国古人混合多种材料制作墨，这些材料包括煮过的兽皮、烧过的兽骨、焦油和煤烟灰。

文身

数百年前，新西兰毛利人用毛毛虫和烧过的树胶制作文身用的墨，用鲨鱼的牙齿作为文身的工具。欧洲人来到新西兰以后，毛利人开始在他们的墨里掺入火药。

便利贴

黏性便利贴有一大优点，即黏性不是太大，可以反复使用。1974年，美国人阿瑟·弗赖伊用一种黏性不是很大的胶水发明了便利贴，而这种胶水是数年前由斯潘塞·西尔弗发明的。

邮票

第一张背面带胶的邮票是1840年发行的英国黑便士，由教师罗兰·希尔发明，他希望寄信更容易也更便宜。今天你如果想买一张原版黑便士邮票，得花300万到400万英镑（相当于人民币2,600万到3,500万元）！

比罗牌圆珠笔

圆珠笔又叫“比罗”，发明于20世纪30年代，名字源于其发明者——匈牙利人拉斯洛·比罗。但圆珠笔真是他发明的吗？并非如此。圆珠笔是美国人约翰·劳德在1888年发明的，他用圆珠笔在皮革上（不再是奶牛身上）写字。不过，他没有因为这项发明得到什么好处。

铅笔

最早的铅笔就是简单的石墨条，是500年前人们从英格兰北部坎布里亚郡山上的矿井里挖出来的。牧羊人用石墨条给他们的羊画标记。

印刷术

印刷术的发明是人类历史上最大的技术进步之一。这意味着你只要学会了阅读，就可以学习书里的任何东西，比如你此刻正在看的这些小知识点。但经历了1,000多年的漫长岁月，印刷术才从它的发源地中国传到欧洲的城市，这让人稍感意外。

木版印刷

中国佛教徒早在公元200年就开始使用木版印刷。他们在木头上雕刻花朵等图案以及文字，然后把雕好的木头印在丝绸上。（就像橡皮图章那样，没有雕的地方会留下墨痕，而雕过的地方不会。）

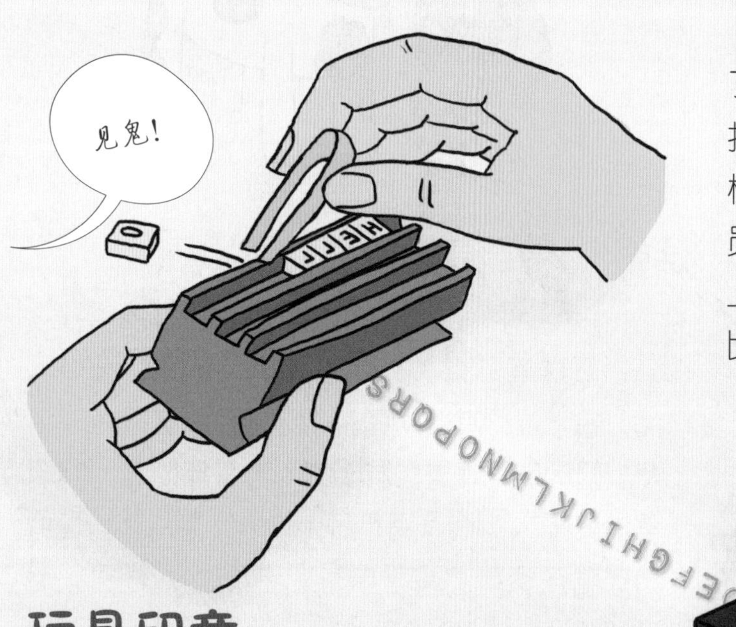

打字机

1829年，美国人威廉·伯特发明了“排字机”，这可能是世界上最早的打字机，但不是很好用。威廉发明这台机器是为了提高文秘人员的工作效率，但实际上，这台机器打起字来比手写还慢！

玩具印章

上图这套“约翰公牛印刷工具”是伦敦卡森·贝克有限责任公司生产的，在20世纪50年代是一种很流行的玩具。每一个小橡胶块上都有一个字母，你可以在卡槽里排列这些字母来组成不同的单词。但问题是：橡胶块会砰的一声弹出去，掉得满地都是，然后被真空吸尘器吸走。

活字印刷

有了活字以后，印刷就更容易了。大约在公元1045年，印刷工毕昇用烧制过的黏土制作中国汉字，他发明的活字印刷术可以印刷数千份东西。

印刷工在金属板上涂一层蜡，然后在蜡上面排列活字，组成词语。
把金属板加热，让蜡软化，然后在汉字上放一块平板，将汉字压进蜡里。
把汉字涂上墨后，将一张纸放到汉字上压一下，让墨印到纸上，然后把纸拿下来。
好哇！只要再印1,285页就可以了……

印刷机

15世纪40年代，约翰内斯·谷登堡将活字印刷术引入德国后，这种印刷术在欧洲流行起来。谷登堡还设计了世界上第一台印刷机。他为教会印刷的早期作品包含一本《圣经》。排列一页《圣经》上的字母大概需要半天的时间，而一本《圣经》有1,286页！他的工人们花了大约三年时间才印刷了180本《圣经》，但这还是比手工抄写方便。

人类如何发送信号？

下次你给别人发短信时，想想生活在计算机时代以前的祖先吧。那时跟人“聊天”需要用到烟雾、镜子、灯光、手势信号等。但18世纪时，电力的崛起带来一系列基于电报技术的新发明。终于，人们只要按按键，就可以立即将信息传到远方了。

这是说你家着火了。

烟雾信号

人类学会控制火以后，很快也学会把湿草放到火上来控制烟了。古代中国人、古希腊人和印第安人常常使用烟雾信号来通报危险。

机械臂也曾用于发送信号，其特殊代码由196种组合方式构成。

臂板信号机

你想找人帮忙时有没有挥手示意过？1794年，法国人克劳德·沙普用这个古老的办法发明了一个臂板信号系统。每个站点接收到一个信号时，就把信号传递到下一个站点。最终，沙普的臂板信号站点网络遍布了整个欧洲。

今天信号不好啊……

日光反射信号器

镜子能反射阳光，特别适合被用为信号发送设备。19世纪20年代，德国测量员将这个想法付诸实践。1869年左右，英国官员亨利·曼斯发明了曼斯日光反射信号器，即一种利用阳光发送信号的装置，信号发送范围达到56千米。

传真机

亚历山大·贝恩是个精明的苏格兰人，他发明了电子时钟，还沿着爱丁堡—格拉斯哥的铁路铺设了电报线。19世纪40年代，他设计了最早的传真机之一，能准确复印文本和图片，把它们像“打电话”一样传给别人。但现在有了电子邮件，很少有人使用传真机了。

贝恩传真机的一些活动部件来自他设计的时钟。

电报机

第一台真正实用的电报机是威廉·库克和查尔斯·惠特斯通在1837年发明的。这种电报机是供新修的铁路使用的，而今这些铁路蜿蜒蛇行，横跨英国。电报机使用了五根电线，向五根指向字母的磁针发送信号。这种电报机虽然易于使用，但价格高昂。因此，他们在1845年又发明了只用一根电线的电报机。

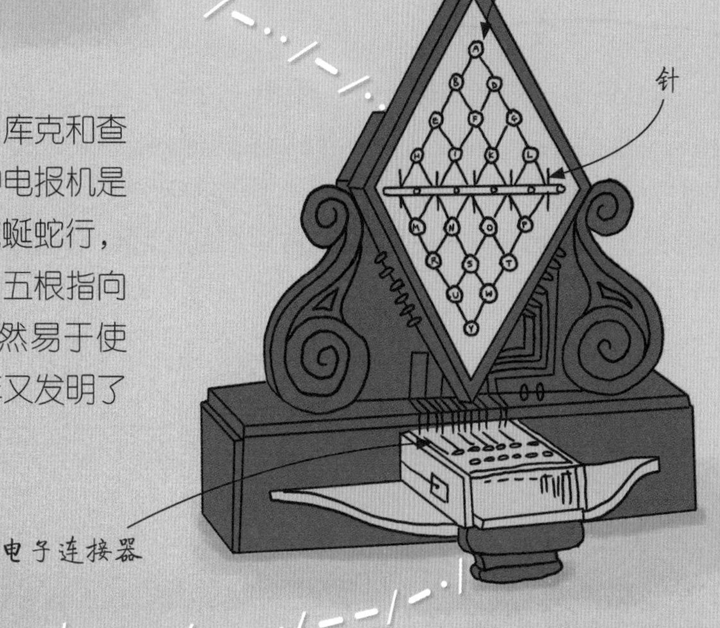

点……点……
点—点—点……
长划……这要
把我整疯了。

莫尔斯电码

莫尔斯电码得名于其发明者美国人塞缪尔·莫尔斯，但约瑟夫·亨利和艾尔弗雷德·韦尔在1836年帮助他完成了发明。你按电报键发送由短信号（点，英文为dots或dits）和长信号（长划，英文为dashes或dahs）组成的代码，另一端接收信号的人将其解码，便可读取信息。

无线电

收音机、电视、计算机、手机、无线对讲机——这些通信工具都使用了无线电波。无线电波很像光波，传播速度也与光波一样快，但我们看不见无线电波。人们掌握了无线电波的原理后，就开始争先恐后地发明新的通信方式。

无线电波

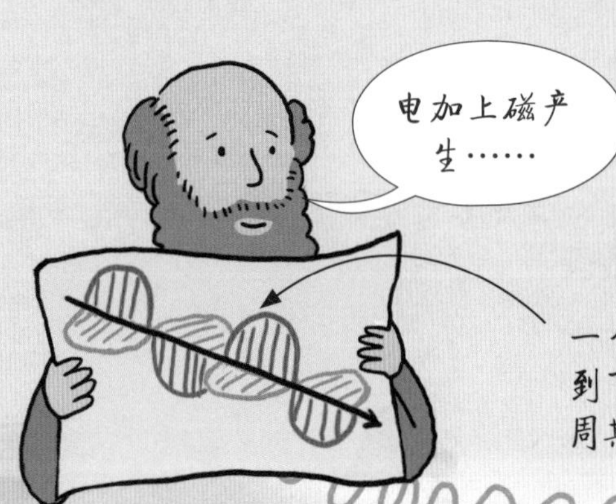

电加上磁产生……

苏格兰人詹姆斯·克拉克·麦克斯韦（左图）是发现无线电波的先锋人物。1864年，他提出磁和电加在一起可以产生不可见的波。

一个完整的“波”（从一个波峰到下一个波峰的距离）叫做一个周期。一个周期的长度是波长。

……无线电波！可算抓到你了！

1885年，德国科学家海因里希·赫兹（右图）用火花间隙当天线，测试了麦克斯韦的理论。当一个火花蹦出火花间隙时，就表示它接收到了无线电波。为了纪念赫兹，一个周期（每秒钟波的振动次数）称作一赫兹（Hz）。

那么是谁发明了收音机呢？我们列出了几个疑似发明者……

特斯拉

塞尔维亚人尼古拉·特斯拉曾在美国为托马斯·爱迪生工作，后来自己单干。他发明了巨大的“特斯拉线圈”来发送和接收无线电波。1892年，他设计了一台收音机，但还没来得及测试，实验室就被烧毁了。因此，马可尼被誉为收音机的发明者。

马可尼

1894到1897年间，年轻的意大利人古列尔莫·马可尼利用发射机和接收器将无线电信号平稳地传播到远处。他去英国向人们展示这套设备，该设备发射的无线电信号穿越了布里斯托尔湾。1901年，他用这套设备发射的信号穿越了大西洋。

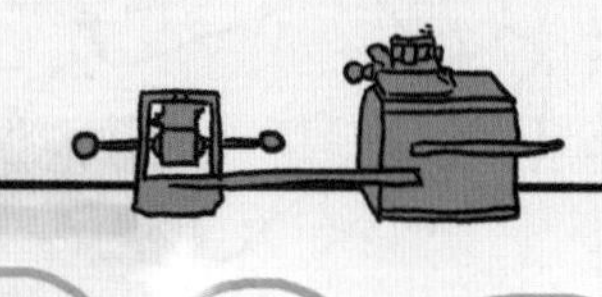

晶体管

1947年，晶体管取代了笨重的真空管。第一根晶体管是在美国的贝尔实验室发明的，看起来像一大团烧煳的意大利面条！现代晶体管非常小，在一平方英寸的电路板上就能安装数百万个晶体管。

最早的晶体管收音机发明于20世纪50年代，是便携式的。

早期的收音机里用的是真空管，会造成收音机发烫！

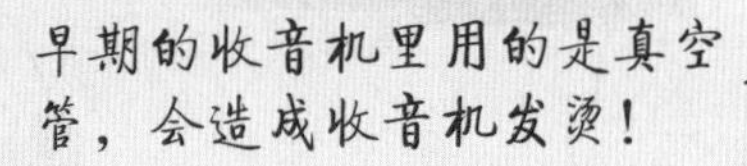

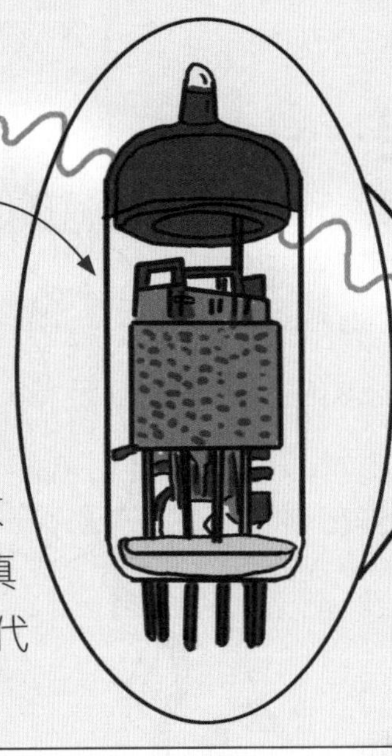

真空管

1904年，英格兰人约翰·弗莱明发明了真空管。1907年，美国人李·德福雷斯特改进了真空管，使收音机能在电流大幅度变化的情况下正常工作。改进后的真空管被用于收音机和电视，从20世纪40年代起，还被用在最早的计算机里。

J. C.博斯

印度人杰加迪斯·钱德拉·博斯是一个全能型天才，曾经写过科幻小说。19世纪90年代，他试验了新的收音设备，启发了其他收音机发明者（如马可尼），但他对发明通信系统不感兴趣。

斯塔布菲尔德

内森·斯塔布菲尔德是一个美国农场主，在19世纪80至90年代发明了一种无线广播系统。1902年，他向人们展示这个系统，将“来自圣诞老人”的信发给了在田野上的几个学生。但这不是真正的收音机，只是运用了磁感应原理。

电话

公认的第一部实用电话问世于1876年，是出生在苏格兰的亚历山大·格雷厄姆·贝尔发明的。但在此后几年里，有将近600人发明了电话，向贝尔发起了挑战。他们这么做是为了钱：如果你能获得世界上最实用的一种发明的专利，就能大赚一笔！所以人们自然不会放过这样的机会。现在苹果公司也为自己的苹果手机申请了专利。以下是一些精彩的故事。

电话发明的真相……

贝尔发明电话前，意大利裔美国人安东尼奥·梅乌奇就发明了一部实用的电话。但他在1871年申请的专利表述不清，贝尔比他更精明。然而，虽然梅乌奇在1889年去世时身无分文，113年后，美国国会宣布贝尔抄袭了他的创意，把属于梅乌奇的荣耀还给了他。

赖斯的电话

约翰·赖斯是德国一所学校的教师，在1861年发明了一部电话。这部电话只能说勉强可用。不过赖斯测试这部电话时说了一句很奇怪的话（见左图）。他还发明了telephon一词，意思是“远处的声音”。

汽车电话

早期的电话当然是通过电缆连到墙上的。（现在很多电话还是这样。）最早的移动电话问世时间要晚得多。右图这种MTA型号的电话是1956年由瑞典的爱立信公司发明的，属于最早的汽车电话之一——但你的车得够大才装得下！

智能手机

第一部智能手机是什么样子的？嗯，想起来了。美国国际商用机器公司（IBM）在1994年推出了西蒙个人通信设备，你可以用它发电子邮件和短信，或者查看日程表……但电池只能用一个小时左右！

苹果手机

苹果手机问世于2007年，是史蒂夫·乔布斯（1955—2011）的智力产物，他是苹果公司的联合创始人。苹果公司的许多员工都为苹果手机的发明做出了贡献，其中就有约翰·凯西。2000年，约翰·凯西结合iPod和电话，发明了一种可打电话的播放器（Telipod）。这就是后来的苹果手机。

可视电话

20世纪60年代中期，贝尔公司推出了可以视频的图像电话（左图），但不是很成功。现在我们有多种可打视频电话的手机应用，比如斯堪的纳维亚人尼克拉斯·詹斯特罗姆和亚努斯·弗里斯在2003年发明的讯佳普。

移动网络

移动电话使用的是“移动网络”，即可以提供网络连接的地面接收站。但有些电话通过绕地球运行的通信卫星进行连接。1962年7月10日，人们打通了第一个卫星电话，用的是当时刚发射的通信卫星1号。

晨鸟号（1965）是最早的通信卫星之一。

移动电话

第一部实用的移动电话是摩托罗拉公司在1973年设计的 DynaTAC。马丁·库珀还从摩托罗拉公司打出一个被载入史册的电话，告诉贝尔实验室的乔尔·恩格尔，摩托罗拉抢在他们前面发明了移动电话。

密码和暗号

许多通信手段都使用了密码，比如臂板信号机、日光反射信号器和莫尔斯电码，利用密码让机械臂、闪光和电码中的点和长划成为可解读的信息。密码还能帮我们保密。自古以来，人们就使用密码防止信息落到敌人手中，破解重要的密码还帮助一些国家赢得了战争。

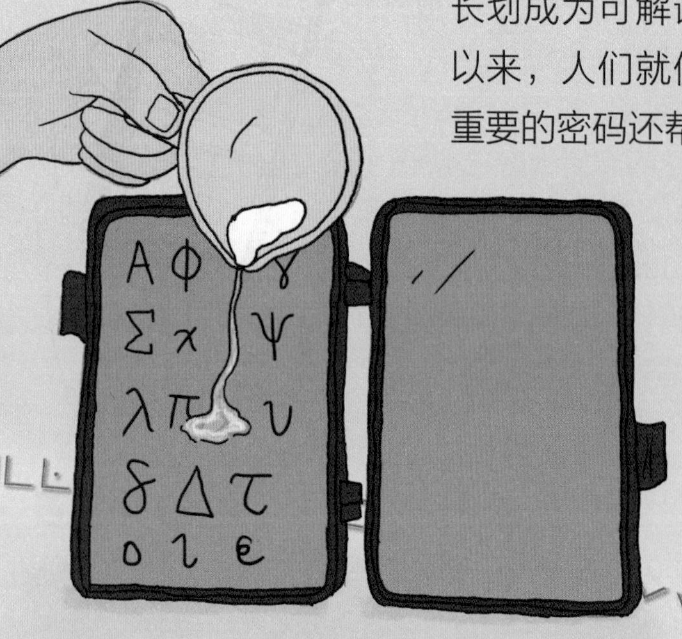

石蜡笔记本

公元前480年，住在波斯的希腊人德马拉托斯想给希腊人通风报信：波斯人正在组建一支大军。于是他用了一本石蜡“笔记本”，将石蜡刮掉，把消息写在里面的木片上，再重新抹上石蜡。笔记本被偷偷带给希腊人，希腊人将石蜡刮掉，找到了木片上的文字。这不是真正的密码，但不失为聪明之举。

信息棍

斯巴达是古希腊的一座城邦，这里的人把信息写在条状的兽皮上，每个字母之间隔得很开。收信人把兽皮卷在一根大小合适的木棍上时，排列出的字母会显示传达的信息。试试这种加密手段吧，你可以把一张纸条卷在一支铅笔或者一根扫帚把上。

吉卜赛暗号

古代旅行者有自己的暗号。他们从一个城镇走到另一个城镇时，会在墙上或者大门上刻上某种标记，告诉其他旅行者哪些人家很友善（或不友善）、哪里有恶狗、哪里容易要到钱等。这种符号语言有时被称为“流浪者的暗号”。

主人在家　乞讨的好地方　上路吧

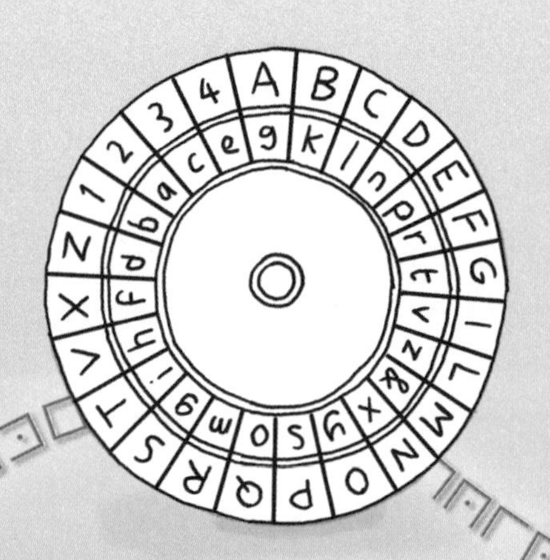

伟大的密码

用于重要场合的密码，比如在战争中使用的密码，破解起来很困难。1626年，安托万·罗西尼奥尔和他的儿子在法国发明了一种密码，用两到三个数字代表一个法语的音节。这种密码被称为“伟大的密码”，保密性非常强，直到1893年才被人破解！

密码盘

15世纪60年代，意大利人莱昂·阿尔贝蒂发明了密码盘。密码盘由两个大小不同的盘组成，两个盘上都有字母表，两套字母表能对齐。外圈的字母也叫明文，用于把信息设置成密码文（内圈的字母）。要想破解，就需要知道应该对齐哪两个字母。

制导系统

在美国，女演员海迪·拉马尔和作曲家乔治·安泰尔为第二次世界大战中使用的鱼雷发明了一种制导系统，使用（自动钢琴上的）钢琴纸卷创造了一种无法破解的密码。这种密码虽然当时没被采用，但后来促进了无线网络和蓝牙的发明。

破解英格玛

第二次世界大战中，纳粹德国的机密信息被一种叫英格玛的机器编成密码。这种编码机是阿瑟·谢尔比乌斯在1918年发明的。波兰人最先破解了英格玛密码，然后把破解方法教给了英国人、法国人和美国人，由此帮助同盟国赢得了战争。

电视和视频

在电视发明早期，发送图像前需要先把它分解成比特，然后将比特传送给接收器，接收器再将比特重新拼成完整的图像。最开始需要使用有活动部件的机械来操作，生成的图像也模糊不清。直到无线电技术的进步促进了电子广播的发明，我们才在屏幕上看到十分清晰的图像。

尼普科圆盘

德国人保罗·尼普科在1884年最先开始分解图像。他用了自己发明的尼普科圆盘。这是一个旋转的盘，上面有排成螺旋形的孔。圆盘将图像投射到一个光电传感器上，光电传感器再将图像转化为或明或暗的“信息”。另一个尼普科圆盘负责破解信息，还原图像。

旋转的圆盘，上面有排成螺旋形的孔

硒电池

光

以同样速度旋转的第二个圆盘

图像

詹金斯电视

在美国，C.弗朗西斯·詹金斯于1923年设计了一种尼普科式的电视，并在1928年创立了美国第一家电视台。弗朗西斯还发明了蜡纸盒（用来装果汁或牛奶）以及飞机弹射器。

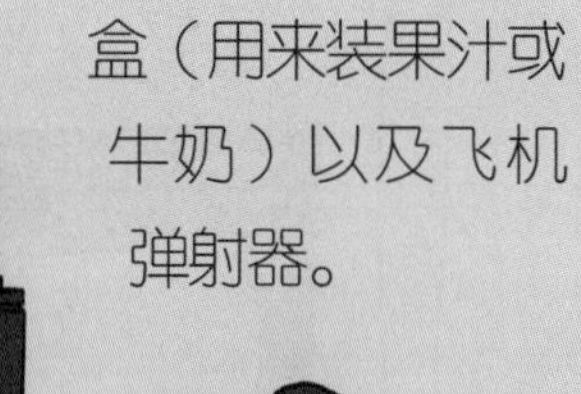

电视之父约翰·洛吉·贝尔德

苏格兰人约翰·洛吉·贝尔德可以说是电视之父。因为太穷，他早期只能用废品发明电视。1928年，他做了首次横跨大西洋——从伦敦到纽约——的电视广播。第二年，他开始为英国广播公司新推出的电视节目做广播。

贝尔德自制第一台电视扫描仪时使用的材料包括：一些尼普科圆盘、一个旧帽盒、织补针、自行车灯和大量胶水。

阴极射线管

以前的电视机体积庞大而沉重，里面有一根阴极射线管（CRT）。在这根阴极射线管里，电子束射在发磷光的表面（即屏幕）上从而产生图像。第一根阴极射线管是德国科学家卡尔·布劳恩在1897年发明的。20世纪20年代，出生在俄罗斯的弗拉基米尔·兹沃里金改进了布劳恩的设计（左图）。阴极射线管发明后，电视就成为电子一族了。

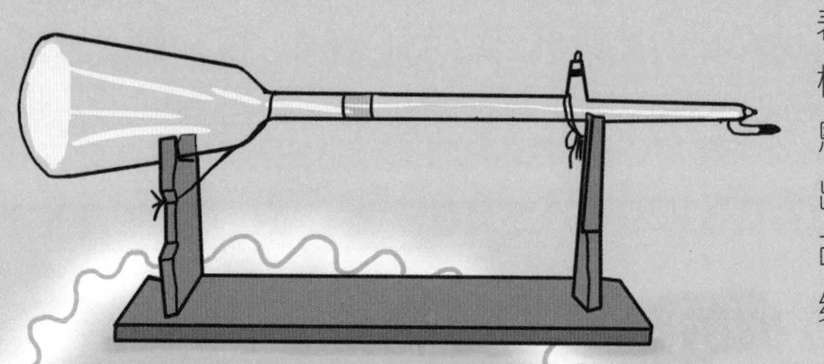

闪光自动化装置

世界上第一个电视遥控器是齐尼思公司在1950年发明的，叫做“懒骨头”。这种遥控器用一根电线与电视机连接。这家公司在1955年发明的“闪光自动化装置”就更先进了，这种遥控器能在电视上打出一道“神奇的光”来换台或者调节音量。（当然，除非你的狗坐在前面挡住了。）

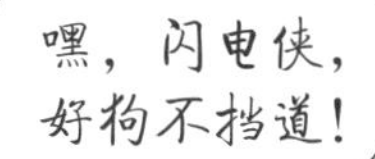

录像机

1995年数字影碟（DVD）发明以前，人们用录像机（VCR）来录音和回放节目。美国歌手宾·克罗斯比想给自己的表演录像，于是在20世纪50年代协助安派克斯公司研发了第一台录像机VR-1000。这台机器有一个煤气炉那么大！

计算机

计算机也许是现代最重要的一项发明。最早的计算机是不到一个世纪前发明的，样子像恐龙，身子大脑袋小。但现在的计算机已经大不相同。仅一部现代智能手机就要比美国航空航天局在1969年使用的所有计算机强大数百万倍。（就是这样的计算机把人类送上了月球！）

计算机的起源：提花机

计算机程序的创意来自250多年前的法国纺织业。如果要在一台提花机上织出花纹，就需要安装一串卡片，每个卡片上都打了一组孔，织机根据孔的样式穿线。

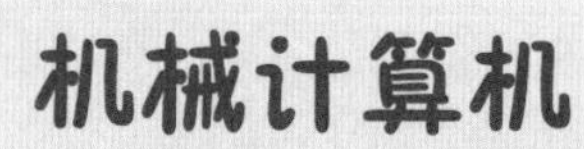

机械计算机

英格兰人查尔斯·巴比奇（1791—1871）设计了最早的机械计算机：“差分机”和“分析机”。它们体型巨大，有成千上万个用于复杂运算的运转部件。但查尔斯没能造出这些计算机。

计算机奇才

埃达·洛夫莱斯（下图）是史上第一个计算机程序员，她是著名诗人拜伦勋爵的女儿，曾在查尔斯·巴比奇手下工作。美国海军少将格雷丝·霍珀（1906—1992）编写了计算机代码，其中包括COBOL语言。据说，她还提出了“computer bug”这一术语，因为当时有一只蛾子掉进了她的计算机。

穿孔制表机

美国人口普查局负责收集美国公民的信息，比如出生及结婚日期。为了更好地处理数据，该局一个名叫赫尔曼·霍利里思的员工从19世纪90年代开始，发明了一系列“制表”机。这类制表机使用打了孔的卡片来保存数据。

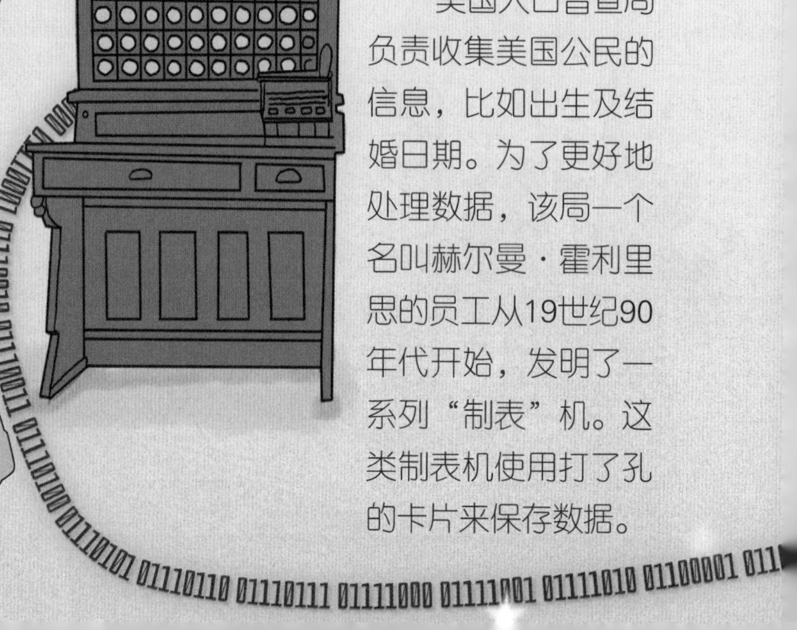

树莓派

今天的集成电路中可能有数以百亿计的晶体管，但体积还是很小。计算机的性能越来越强大，体积却越来越小。所以你能用便宜的微型处理器，比如2005年推出的Arduino和2012年推出的树莓派，完成你感兴趣的项目。

第一只鼠标

1967年，道格·恩格尔巴特和比尔·英格利希在美国斯坦福研究所发明了世界上第一只鼠标。道格负责设计，比尔负责制作。这只鼠标有一个木头外壳和两只金属轮子。

芯片

计算机的体积需要缩小才能供普通人使用，这可以通过集成电路（也就是芯片）实现。1958年，杰克·基尔比在得克萨斯仪器公司发明了第一块集成电路。后来，英特尔公司的芯片促成了最早的个人计算机的发明，比如牵牛星8800和1976年生产的第一代苹果电脑。

巨人计算机1号

巨人计算机1号是英国工程师汤米·弗劳尔斯在1943年设计的，建造这台计算机是为了破解二战时期德国的密码。这台计算机重约一吨，里面有1,600个真空管，是第一台可以编程的电子数码计算机。

因特网

仅仅在几十年前，因特网还有点像俱乐部，只对一些特定的人群开放，比如科学家和政府工作人员。现在，人们上网就跟接电话一样容易。每天都有十亿多人使用脸书；每分钟都有300多小时时长的视频被上传到YouTube视频网站上。但这一切是从哪里开始的呢？

阿帕网

因特网最初的原型是阿帕网。阿帕网是1969年专门为美国国防部高级研究计划局设计的。这个由发明家组成的美国政府部门想让一些特定的群体交流想法。交流的方式之一就是电子邮件，这个电子邮件系统叫做@ARPANET，是雷·汤姆林森在1971年发明的。

分组交换

因特网通过分组交换进行通信。这是一种通过共享路径发送小的电子数据“包”的方式，在20世纪60年代由在美国的伦纳德·克兰罗克和保罗·巴兰以及在英国的唐纳德·戴维斯发明。1973年，文特·瑟夫和鲍勃·卡恩在美国写出了传输控制协议，即分组交换“规则”。也是在那一年，阿帕网成为“因特网”，走向了世界。

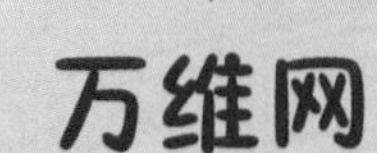

万维网

万维网（WWW）就是因特网上的所有信息，是英格兰计算机科学家蒂姆·伯纳斯－李发明的。20世纪80年代，蒂姆在欧洲核子研究组织工作。他发明了HTML（超文本标记语言）。欧洲核子研究组织开始在一个网页上使用这种语言，这个网页成为世界上第一个网页。

Instagram

2010年，凯文·希斯特罗姆和迈克·克里格设计出Instagram——一款照片和视频分享应用。两年以后，脸书以大约十亿美元的价格收购了这款软件。

脸书

社交网站脸书是美国学生马克·扎克伯格在2004年设计的。他最初将这个网站取名为Facemash，但好在他改变了主意。四年以后，他成了全球有史以来最年轻的亿万富翁。

射频识别

物联网使用了射频识别（RFID）技术。信用卡、商店里的防盗标签、宠物和农场牲畜体内的芯片都使用了这项技术，通过电磁场实现连接。射频识别技术的发明是很多人的功劳。比如1945年，俄罗斯发明家莱昂·塞里明把它用在一个窃听装置里来监视美国人。

物联网

将来有一天，你家里的汽车、冰箱、计算机，也许还有你的宠物，都会联网，这个网叫做物联网（IoT），已经投入使用了。1982年，第一件连接物联网的“东西”是美国卡内基梅隆大学的一台软饮售货机。学生可以上网查这个售货机里是否有冰镇饮料，就不用亲自跑一趟去看了。

稀奇古怪的发明

英格兰杰出的科学家斯蒂芬·霍金曾说：“人类最伟大的成就是通过沟通取得的；而最大的失败是不沟通造成的。”确实如此，但交谈并非总那么简单，看看以下一些关于沟通的奇葩发明就知道了。

给鬼打电话

据说在1920年，美国最著名的发明家托马斯·爱迪生提出要发明跟鬼魂交流的“通灵电话”。这个故事很可能是杜撰的，但是很有趣。

贝尔德的创意……

约翰·洛吉·贝尔德有一些成功的发明，比如电视和保暖袜，但也有一些失败的发明，比如：

- 玻璃剃须刀刀片。他想做一把不会生锈的剃刀，但问题是这样的剃刀会碎。好疼！
- 人造钻石。钻石是一种碳，石墨（铅笔的“铅”）也是一种碳。但钻石是在高压高温条件下，在地下天然形成的。贝尔德认为，他可以通过加热石墨制造钻石。但他造出来的不是钻石而是麻烦：他做实验时切断了格拉斯哥的电源，让整个城市漆黑一片。

烫手的网球

1953年，美国发明家马尔温·米德尔马克因为发明了兔耳电视天线成了百万富翁。他还尝试过做水力土豆削皮机以及把网球放到微波炉里加热来恢复网球的性能，但都失败了。

狗语翻译机

想跟你家的狗聊天吗？2002年，日本一个研究小组发明了一台机器。这台机器可以听狗叫，并根据狗叫声破译出狗狗的情绪（比如开心或难过），然后显示在屏幕上。现在世界上已有各种版本的狗语翻译机，能听懂不同狗狗的方言。

猫爪感应器

猫会在键盘上乱走，把你的家庭作业变成天书，这事你知道吗？这款计算机软件可以拯救你。只要监测到“猫式输入法”，软件就会发出警报声（猫讨厌这样的声音，听了就跑），并“锁定”键盘。

坎皮略的疯狂电报机

早期的电报机形式多样，其中一些脑洞特别大。1804年，西班牙科学家弗朗西斯科·萨尔瓦·坎皮略提议，把一群人连上电线，每个人代表一个不同的字母或数字。通过电线传递的信息电击他们时，他们会大声叫出自己代表的字母，从而拼写出信息。好可怕！

不靠谱的预测

当我们回顾历史时，很容易嘲笑过去的人们有过的想法和说过的话，觉得匪夷所思。当然人们当时并不知道未来会发生什么，所以其实不是他们的问题。但下面这些话或许还是会把你逗笑。

电影的未来

“去电影院看电影不过是一时狂热。”

——查理·卓别林，演员及电影导演，1916年

有声电影的未来

“难道有谁会想听演员说话吗？”

——哈里·华纳，华纳兄弟电影公司总裁，1925年

（当然，今天99.999999999%的电影都是“有声电影”。）

电话的未来

“美国人需要电话，但我们不需要，我们有许多信童。”

——威廉·普里斯，英国邮政局总工程师，1879年

（今天，英国有3,300多万家庭安装了有线电话，93%的成年人有手机。）

战争的未来

“无线电时代的到来会消灭战争，因为无线电会使战争失去意义。”

——古列尔莫·马可尼，收音机创始人，1912年

（听起来很可笑，但悲哀的是，无线电只是成了人们发动战争的帮凶。）

计算机的未来

“全球市场对计算机的需求量可能只有五台。”

——托马斯·沃森，计算机公司IBM的老板，1943年

（今天，全世界有20多亿台个人电脑。）

“未来的计算机重量可能不超过1.5吨。”

——《大众机械》杂志，1949年

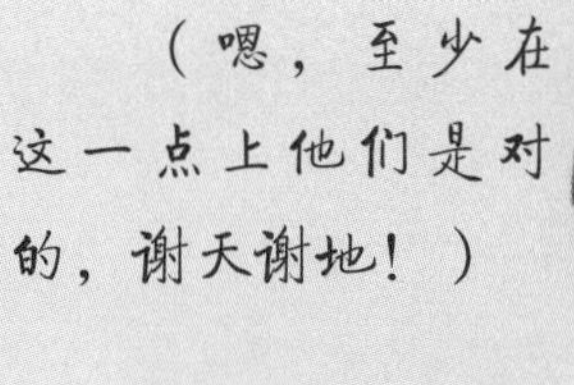

（嗯，至少在这一点上他们是对的，谢天谢地！）

因特网的未来

“我预测，因特网很快就会成为超新星，然后在1996年遭遇毁灭性的失败。”

——罗伯特·梅特卡夫，计算机网络公司3Com的创始人，1995年

（罗伯特承诺，如果他说错了就把自己说的话吃了。现在我们知道，他确实说错了！两年后，他把自己的预测打印出来，加水放到厨房搅拌器里搅烂，然后喝掉了。）

索引

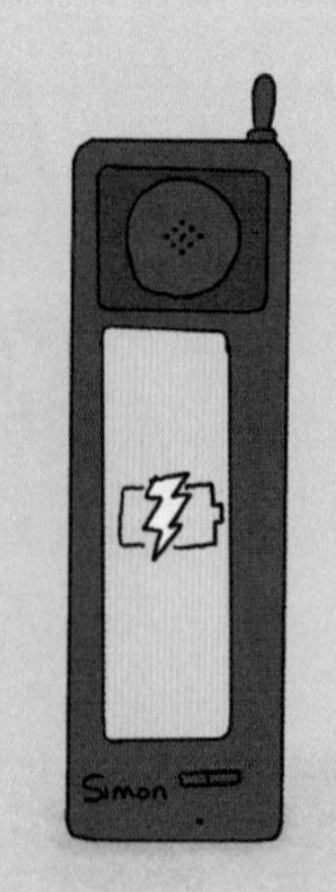

作者简介

马特·特纳出生于英国，20世纪80年代毕业于拉夫伯勒大学艺术学院，毕业后一直担任图片研究员、编辑和作者。他的书题材广泛，涉及博物学、地球科学和铁路等，并为百科全书和分册出版的丛书写过数百篇文章，从大象到抽象艺术无所不包。他现在和家人住在新西兰奥克兰附近，他还是当地海岸警卫队的志愿者，平时也涉猎工艺品的制作。

绘者简介

萨拉·康纳生活在英格兰美丽的乡村，栖身于一座可爱的小木屋里，有几只狗和一只猫做伴。她平时用画笔描绘身边的世界。她总能从大自然中获得灵感，这对她的很多作品都有影响。萨拉之前用钢笔和颜料画插图，但近些年，她开始在电脑上绘画，因为这更适应今天的产业发展。然而，她还是喜欢时不时拿出水彩颜料来，画一画自己花园里的鲜花！

we're
ON A ROLL!

1,000,000,000

Printing words (so you can read this book, for instance) originated with the Chinese. But first, alphabets had to be created. Follow this and many other stories about the methods and machines we use to communicate by signals, code and telephone, radio, TV and computer. You'll find out

- who really invented the telephone
- which code took 267 years to crack
- the inventor of TV built his early machines from junk (it's all he had)
- the idea for the computer program came from cloth weaving

and so much more. And there are wacky inventions (like the machine that translates your dog's barks), and wayward predictions – um, the end of the Internet was going to happen in 1996... (The guy ate his words, literally.)

German schoolteacher Johann Reis invented a telephone in 1861. To see what the horse is up to, turn to page 18.

CONTENTS

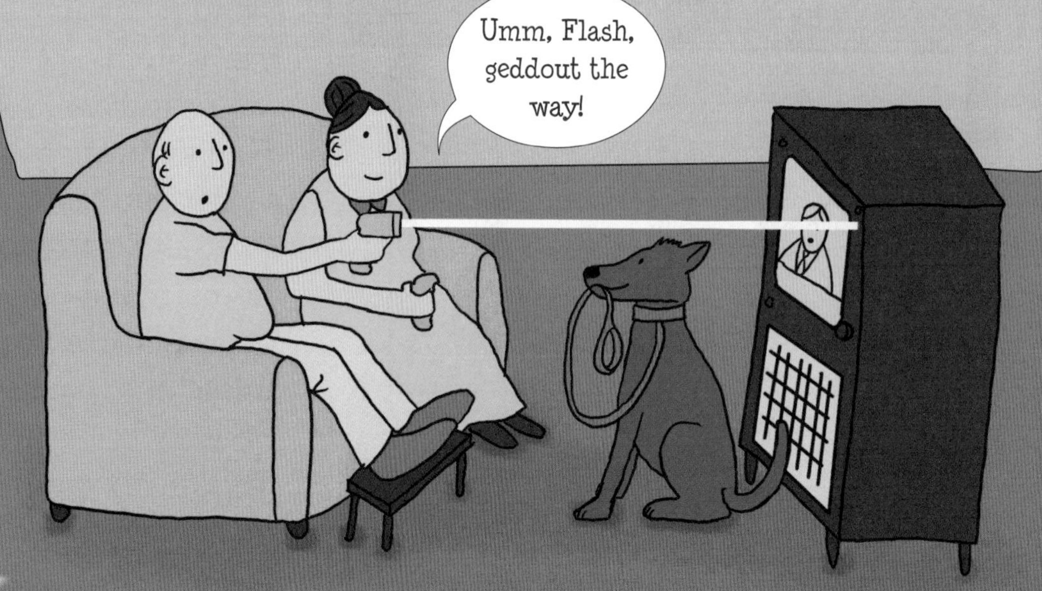

Umm, Flash, geddout the way!

IN TOUCH

The history of communication takes us from earliest times to the invention of alphabets and writing, the printing press, the use of electricity in telegraph and telephone, right through to our modern digital age with its computers, satellites and smartphones.

iPhone, 2007

Our earliest ancestors made stunning drawings on cave walls, so maybe they spoke just as beautifully as they sat around the fire at night. We'll never know, because we have no record of their languages – alphabets had not yet been written. Even into modern times, some cultures didn't bother with writing. The Māori of New Zealand, for instance, only began writing down words in the 1820s, after Europeans visited them. But their traditional tattoos and carvings are a kind of language in themselves.

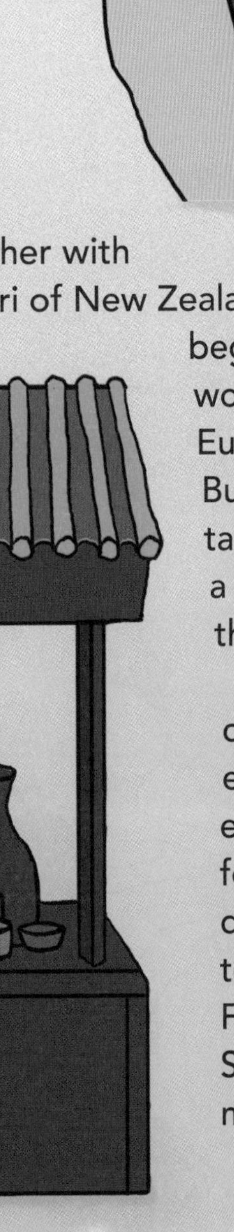

Paper, 105 AD

Today, communicating is easier and faster than ever, and it takes many forms. In any single day we might use telephone, email, text, Facebook, Twitter, Skype, Instagram and more. We sometimes

text other family members even if they're in the same room! In a peculiar way, we've come full-circle back to cave days, speaking directly (electronically) to our friends – even those in other countries – rather than writing them a letter.

Here we look at how those big changes happened. Some were slow, like the gradual, century-by-century changes to ancient picture alphabets. Others have been quick: the Internet, for instance, is less than 50 years old, and Facebook is still a teenager.

Find out who invented paper and pens, printing, email and Instagram. And have fun with some hilarious predictions made by stick-in-the-muds who couldn't see any future for radio, television, cinema, computer...

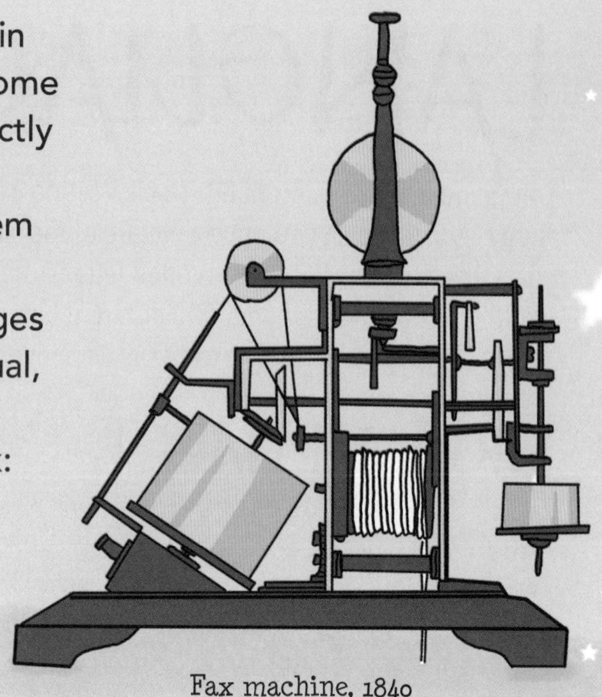

Fax machine, 1840

Vacuum tube (in radio), 1904

Video cassette recorder, 1950s

LANGUAGE

Language is as old as humanity itself. Writing originally came from art: if you wanted to say 'sun' or 'bear', you drew a picture of the sun or a bear. Gradually, down the years, these drawings became coded letter systems: alphabets. Today, around the world, there are more than 6,000 different languages. But except for the most recent languages, we simply can't say who invented them.

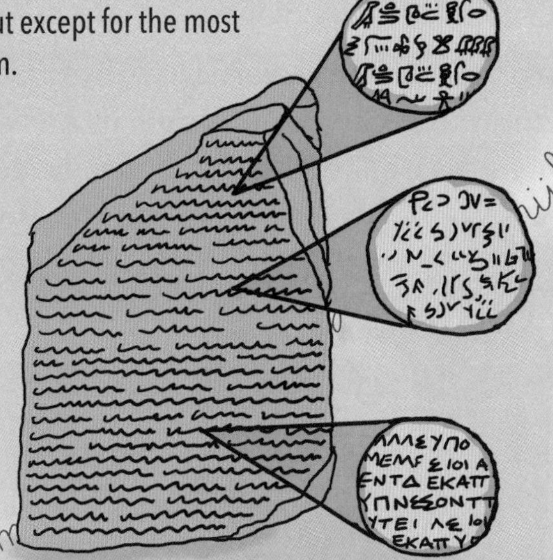

FIRST ALPHABET

People first used alphabets about 5,200 years ago in the Middle East, mainly for writing about taxes and trading. This Sumerian scribe is using a reed to punch wedge shapes in wet clay (which later dried). Sumerian wedge-letter writing is called cuneiform.

ROSETTA STONE

Ancient Egyptian priests used pictograms we call hieroglyphs ('holy carvings'). For ages, we didn't know what these meant. But in 1799, a French soldier in Egypt discovered an old stone tablet carrying the same message in three languages. At last we could work out what those priests were scribbling.

Original pictogram	Later pictogram	Assyrian cuneiform	Meaning
			Mountains
			Fish
			Ox
			Grain
			To Go

PICTOGRAMS

The very earliest writers used pictograms – shapes describing things. So, a pictogram for 'water' was often a wavy line, and 'ox' was a horned head. Later, people began using simpler shapes, which were quicker to draw than pictograms. This table shows some early writing systems from the ancient Middle East.

PITMAN SHORTHAND

Fast-forward to 1830s England, where a teacher, Isaac Pitman, believed that people spent too long writing. 'Time saved is life gained,' he said. Too right! Isaac invented a shorthand language of simple lines and squiggles. Each shape stood for a sound or for a simple word (like 'the' or 'you'). Pitman shorthand is still popular, saving people time today.

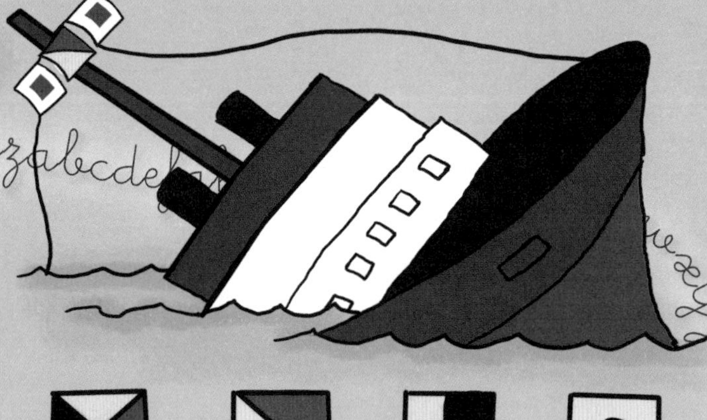

SIGNALS

Before radio, sailors hung flags, like laundry, on their ships' masts to send signals. Captain Fred Marryat devised a basic flag code in 1817, and 40 years later it grew into the International Code of Signals, with 26 flags. You can send a message with a single flag: the A flag means 'I have a diver down.' Or you can string lots together to make words.

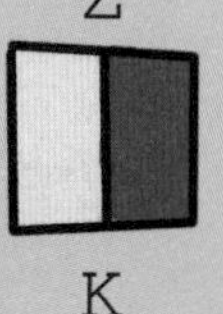

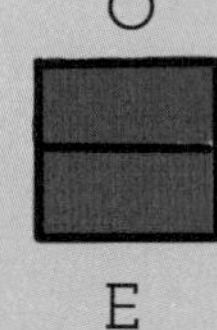

ESPERANTO

Esperanto is a language invented in 1887 by Polish doctor Ludwig Zamenhof. He wanted to make it so easy that anyone could use it: a kind of international language. Currently, more than two million people can speak Esperanto.

WRITING IT DOWN

What is written on is as important as writing. You can't carry stone tablets to the shops! Early people made writing material from plant fibres, and parchment from animal skins. Paper came much later, and pencils and pens are really very recent.

SYMBOLS

The very earliest writing yet found is a group of symbols found at Jiahu in China. They were carved onto tortoiseshells and bones about 8,500 years ago. No one's quite sure what the symbols mean. Magic spells, perhaps.

PAPER

The ancient Egyptians mashed strips of papyrus, a marsh plant, into flat sheets for writing on, and so gave us the word 'paper'. Around 105 AD in China, Cai Lun made paper from rags, bark and old nets.

INK

For ink, early Chinese used a mixture of boiled animal skin, burnt bones, burnt tar and soot.

TATTOOING

Centuries ago, the Māori of New Zealand made tattoo ink from caterpillars and burnt tree gum, and used sharks' teeth as tattoo tools. When Europeans arrived, Māori began adding gunpowder to their ink.

POST-IT

The great thing about sticky Post-It notes is that they're not too sticky, so you can reuse them lots of times. American Arthur Fry invented the Post-It note in 1974, using a not-too-sticky glue invented a few years earlier by Spencer Silver.

STAMPS

The first glue-backed stamp was the British Penny Black of 1840. Rowland Hill, a teacher, introduced it to make letters easier and cheaper to send. To buy an original Penny Black today, you'll need £3-4,000,000 (¥26-35,000,000)!

BIRO

The ballpoint pen is called the 'biro' after Hungarian László Biró, its inventor in the 1930s. Or was he? In fact it was American John Loud's idea, in 1888. He used his ballpoint for writing on leather (when no longer on the cow), but he never benefited from his invention.

PENCILS

The first pencils were simple sticks of graphite dug from mines in the hills of Cumbria, northern England, as much as 500 years ago. Shepherds used them to identify their sheep.

PRINTING

The invention of printing was one of the biggest technological leaps in human history. It meant that if you learned to read, you could learn anything written in books. Like these little facts. So it's slightly surprising that printing took so long – more than 1,000 years – to spread from its birthplace in China to the cities of Europe.

WOOD-BLOCK PRINTING

Chinese Buddhist monks began wood-block printing as early as 200 AD. They carved pictures, such as flowers, and text into wood, then used the carvings to print on silk. (Like a rubber stamp, the uncut parts made inky marks; the cutaway parts did not.)

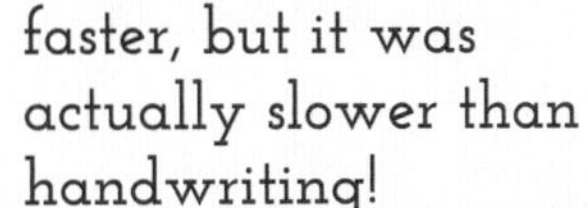

TYPEWRITER

American William Burt came up with the 'typographer' – possibly the world's first typewriter – in 1829, but it didn't work terribly well. He'd invented it to make secretaries work faster, but it was actually slower than handwriting!

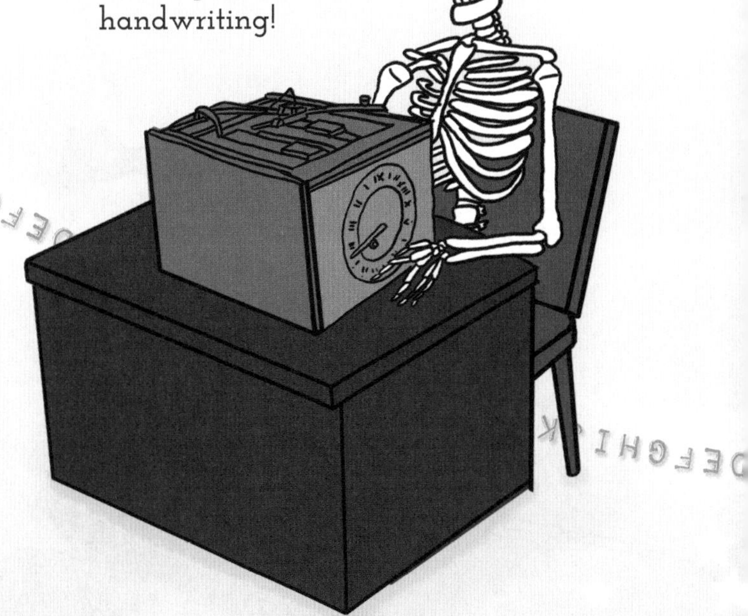

TOY PRINTMAKER

This 'John Bull Printing Outfit', made by Carson Baker Ltd of London, was a popular toy in the 1950s. Each letter was on a little rubber block, and you arranged them in slots to form words. Trouble was, they pinged off all over the place, then went into the vacuum cleaner.

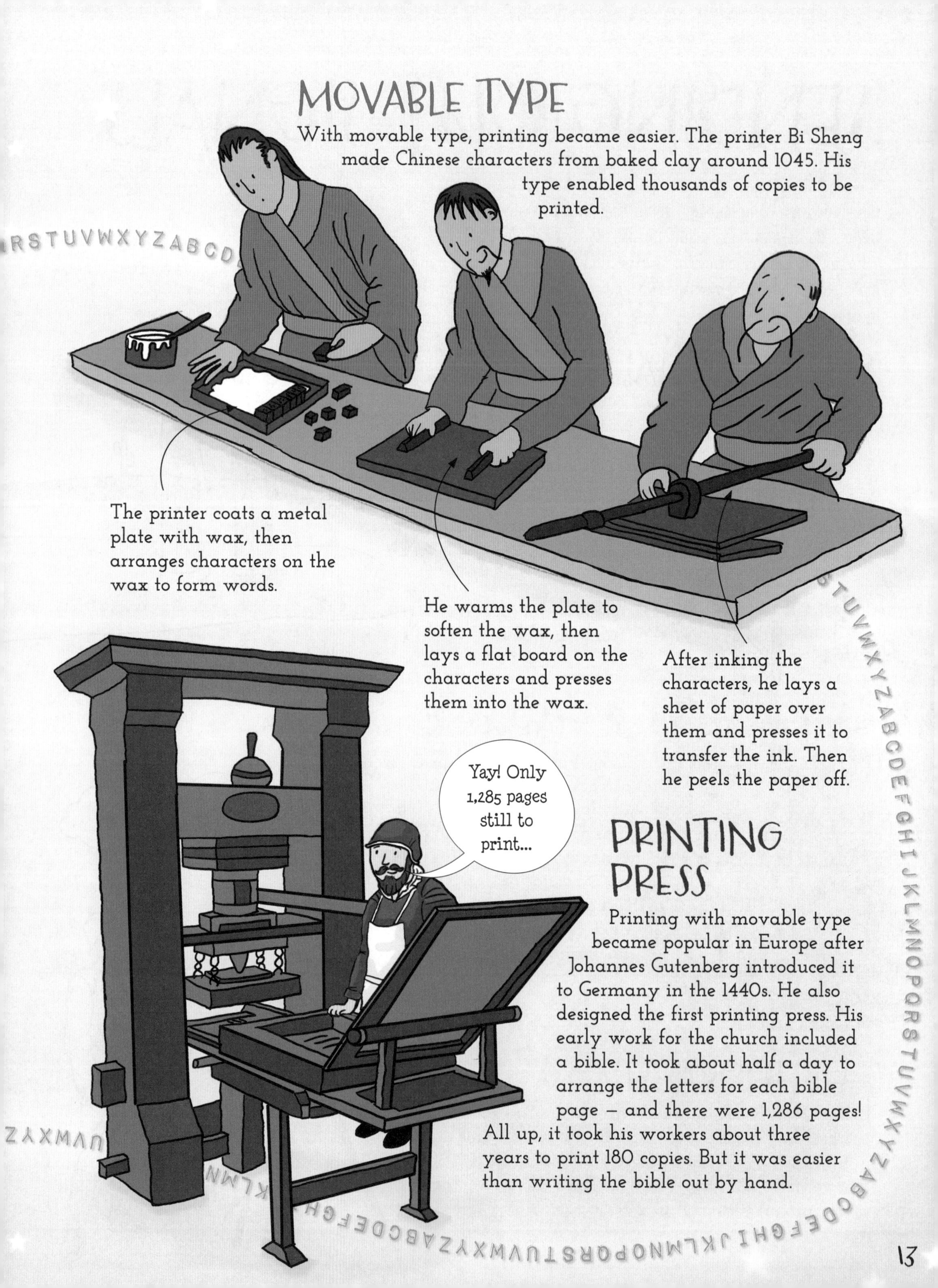

MOVABLE TYPE

With movable type, printing became easier. The printer Bi Sheng made Chinese characters from baked clay around 1045. His type enabled thousands of copies to be printed.

The printer coats a metal plate with wax, then arranges characters on the wax to form words.

He warms the plate to soften the wax, then lays a flat board on the characters and presses them into the wax.

After inking the characters, he lays a sheet of paper over them and presses it to transfer the ink. Then he peels the paper off.

Yay! Only 1,285 pages still to print...

PRINTING PRESS

Printing with movable type became popular in Europe after Johannes Gutenberg introduced it to Germany in the 1440s. He also designed the first printing press. His early work for the church included a bible. It took about half a day to arrange the letters for each bible page – and there were 1,286 pages! All up, it took his workers about three years to print 180 copies. But it was easier than writing the bible out by hand.

SENDING OUT SIGNALS

The next time you text someone, spare a thought for your ancestors from before the computer age. Doing 'chat' involved things like smoke, mirrors, lights, arm signals... But the rise of electric power in the 18th century brought a wave of new inventions based on the telegraph. At last, they could send messages instantly over long distances, simply by pressing keys.

It says your home is on fire.

SMOKE SIGNALS

As soon as humans learnt to control fire, they could also control smoke, by placing damp grass on a fire. Ancient Chinese and Greeks, as well as Native Americans, used smoke signals, often to warn of danger.

Mechanical arms signalled using a special code with 196 combinations.

SEMAPHORE

Ever waved your arms to attract help? Frenchman Claude Chappe put this old trick to use when he invented a semaphore system in 1794. As each station received a signal, it passed it on to the next. Chappe's network of stations eventually spanned all of France.

HELIOGRAPH

A mirror, reflecting sunlight, makes an excellent signalling device. German surveyors used this idea in the 1820s. About 1869, Henry Mance, a British official, invented the Mance heliograph, a sun-signalling instrument with a range of 56 km.

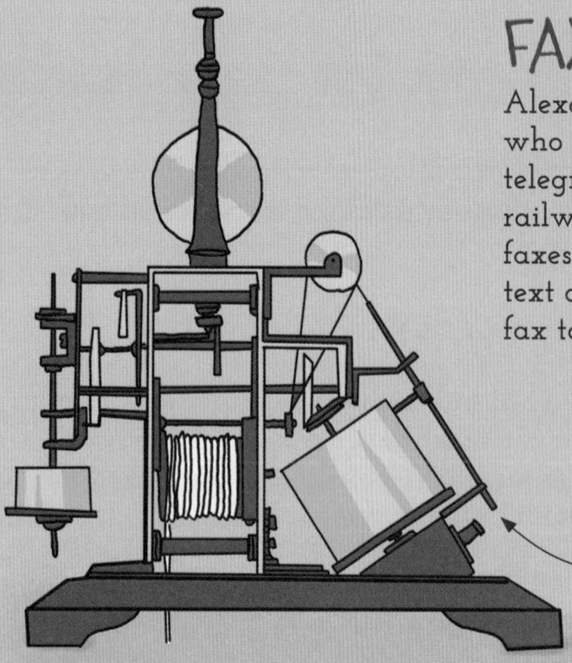

FAX MACHINE

Alexander Bain was a canny Scotsman who invented the electric clock and also laid telegraph lines along the Edinburgh–Glasgow railway. In the 1840s, he devised one of the first faxes: machines that 'telephone' exact copies of text and pictures to people. Few people use the fax today, now that we have email.

Some of the moving parts for Bain's fax machine came from his clock designs.

Letter

Needle

TELEGRAPH

The first really practical telegraph was invented in 1837 by William Cooke and Charles Wheatstone, for use on the new railways that are now snaking across Britain. It used five wires, sending signals to five magnetic needles, which pointed to letters. Though easy to use, it was expensive, so they came up with a single-wire telegraph in 1845.

Electrical connectors

'Dit... dit... dit-dit-dit... dah... this is driving me dotty.

MORSE CODE

Morse code is named after American Samuel Morse, but others – Joseph Henry and Alfred Vail – helped with his 1836 invention. You used the electrical tapper to send a code of short signals (dots, or 'dits') and long (dashes, or 'dahs'), and the person at the other end decoded it to read the message.

RADIO

Radio, TV, computer, mobile phone, walkie-talkie – all these communication tools use radio waves. These are rather like light waves but are invisible, though just as fast. Once radio waves were understood, the race was on to invent new ways of communicating.

Electricity plus magnetism make...

RADIO WAVES

The discovery of radio waves began with Scotsman James Clerk Maxwell (left). In 1864, he suggested that magnetism and electricity working together could create invisible waves.

One full 'wave' (from one crest to the next) is called a cycle. The length of one cycle is the wavelength.

...radio waves! Now we've got you!

In 1885, German scientist Heinrich Hertz (right) tested Maxwell's theory, using a spark gap as an antenna. When a spark jumped the gap, it showed it had picked up radio waves. The cycle (waves per second) was named a hertz (Hz), in Heinrich's honour.

So who invented radio? We've lined up some suspects...

TESLA

Serbian Nikola Tesla worked in the USA for Thomas Edison, then on his own. He invented giant 'Tesla coils' to send and receive radio waves. In 1892, he designed a radio, but his lab burnt down before he could test it. So Marconi is credited as the inventor of radio.

MARCONI

In 1894–97, a young Italian, Guglielmo Marconi, used transmitters ('senders') and receivers ('getters') to send radio signals over steadily greater distances. He went to Britain and demonstrated his kit by transmitting across the Bristol Channel. In 1901, he sent a message over the Atlantic.

TRANSISTOR

The bulky vacuum tube was replaced in 1947 by the transistor. The first transistor, designed at Bell Labs in the US, looked like a tangle of burnt spaghetti, and it was big! Modern transistors are so tiny you can fit millions into 1 sq in of circuitboard.

The first transistor radios of the 1950s were portable.

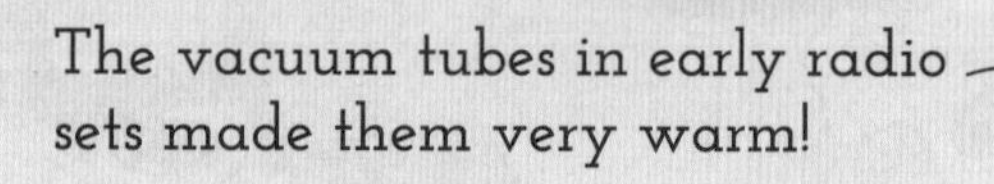

The vacuum tubes in early radio sets made them very warm!

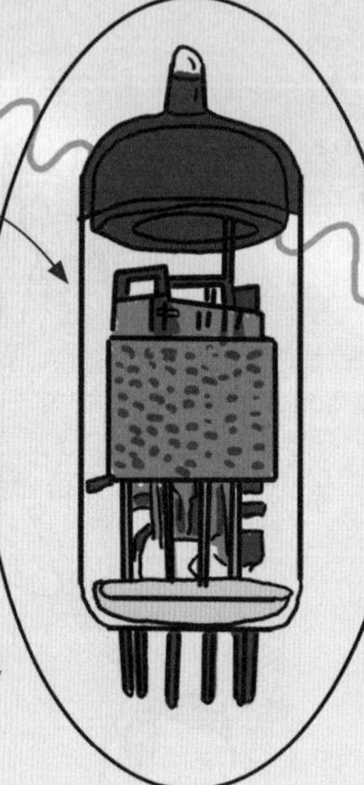

VACUUM TUBE

The vacuum tube, invented by Englishman John Fleming in 1904, and improved on by American Lee de Forest in 1907, allowed for big changes in current. It was used in radio, TV and, from the 1940s, the first computers.

J. C. BOSE

Jagadis Chandra Bose of India was an all-round genius who also wrote sci-fi novels. His experiments with new equipment in the 1890s helped other radio pioneers (like Marconi), but he wasn't interested in making a system of communication.

STUBBLEFIELD

Nathan Stubblefield, an American farmer, invented a kind of wireless broadcasting system in the 1880s–90s. In 1902, he demonstrated it by sending a message 'from Santa Claus' to some schoolkids in a field. But it wasn't true radio; it used magnetic induction.

TELEPHONE

Scottish-born Alexander Graham Bell usually gets the credit for inventing the first workable telephone in 1876, but over the next few years nearly 600 people challenged him with inventions of their own. The reason was money: if you could obtain a patent for one of the world's most useful inventions, you could make a lot of cash! And of course, people have done that, all the way up to Apple with its iPhone. Here are some of the highlights.

THE TRUTH...

Antonio Meucci, an Italian-American, invented a practical telephone before Bell, but his 1871 patent was unclearly worded, and Bell outsmarted him. Although Meucci died penniless in 1889, honour was restored 113 years later, when the US Congress said that Bell had stolen his ideas.

The horse does not eat cucumber salad.

How'd you get my number?

ANOTHER VOICE

Johann Reis, a German schoolteacher, invented a telephone in 1861. It wasn't very good, but it worked, although Reis used an odd message to test it (see left). He even invented the name 'telephon', which means 'far voice'.

CAR PHONE

Early phones were, of course, connected to the wall by electric cable (and many still are). The first mobile phones came much later. This MTA model, made in 1956 by Swedish company Ericsson, was one of the first car phones – but you needed a big car!

SMARTPHONE

The first smartphone? Hmmm, well. The Simon Personal Assistant, launched by IBM in 1994, enabled you to send emails and texts, or check your calendar... but the battery lasted only about an hour!

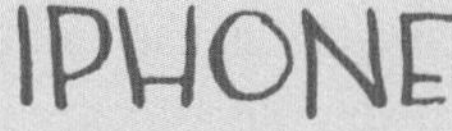

IPHONE

The iPhone arrived in 2007, and was the brainchild of Steve Jobs (1955–2011), co-founder of Apple. Many Apple workers helped invent it. One was John Casey, who created the 'Telipod', from the iPod and a telephone, in 2000. This became the iPhone.

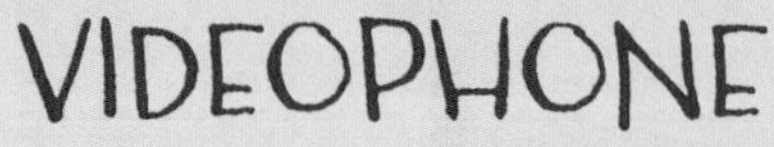

VIDEOPHONE

In the mid-1960s, the Bell company launched this Picturephone, for making video calls, but it wasn't very successful. Today, we have apps such as Skype, invented in 2003 by Scandinavians Niklas Zennström and Janus Friis.

MOBILE NETWORKS

Mobile phones use 'mobile networks': ground stations that provide a connection. Some phones connect instead through the communications satellites orbiting Earth. The first satellite phone call was made on 10 July 1962, using the newly launched Telstar 1 satellite.

Early Bird (1965) was one of the first communications satellites (comsats).

MOBILE PHONE

The first practical mobile phone was this Motorola DynaTAC of 1973. Martin Cooper at Motorola famously phoned Joel Engel at Bell Labs to tell him they'd beaten Bell to an invention.

CODE

Lots of communication methods use codes – the semaphore, heliograph and Morse tapper used code to make their mechanical arms, flashes and dit-dahs readable. But codes can also provide secrecy. Since ancient times, we've used code to keep messages out of enemy hands, and cracking important codes has helped countries win wars.

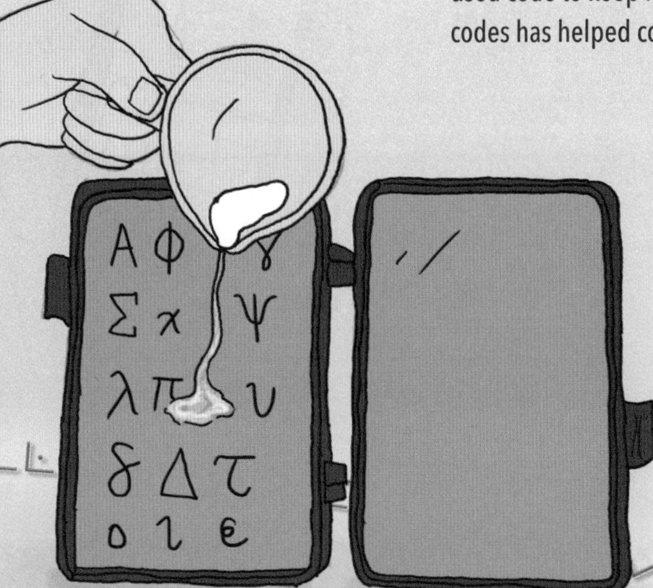

HIDDEN MESSAGE

Demaratus, a Greek living in Persia in 480 BC, wanted to warn the Greeks that the Persians were building an army. So he took a wax 'notebook', scraped off the wax, and wrote his message on the wood base. Then he rewaxed it. This was smuggled to the Greeks, who scraped off the wax to find the message. Not really code, but still clever.

MESSAGE STICK

The people of Sparta, an ancient Greek city-state, wrote messages on strips of animal hide. They spaced the letters widely. When the recipient wrapped the strip around a wooden pole of the correct size, the letters lined up to reveal the message. You can do this code trick with a strip of paper around a pencil or a broom-handle.

GYPSY CODE

Travellers in olden times used their own secret code. Wandering from town to town, they'd scratch marks on walls or gates to tell other travellers about friendly (or unfriendly) householders, a dangerous dog, easy pickings and so on. Their sign language is sometimes known as 'hobo code'.

= Owner is in

= Good place for a handout

 = Hit the road

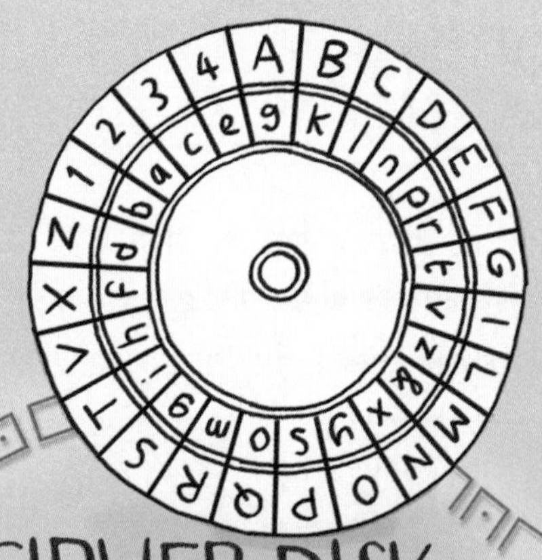

CIPHER DISK

This cipher disk was invented by Italian Leon Alberti in the 1460s. It has two discs of different sizes, each with an alphabet, and the two sets of letters can be lined up. The outer ring, or plaintext, is used to set messages into ciphertext (the inner ring). Decoding depends on knowing which two letters to line up.

THE GREAT CIPHER

Codes used seriously – in war, for instance – can be very hard to break. In 1626 in France, Antoine Rossignol and his son invented a code in which each French syllable was represented by two or three numbers. Their code was known as 'The Great Cipher'. It was so strong no one cracked it until 1893!

Where are the French hiding?
Um... 34 - 561 - 192, Sir...

We're
ON A ROLL!

GUIDANCE SYSTEM

In the US, actress Hedy Lamarr and composer George Antheil invented a guidance system for torpedoes in World War II. It used a piano roll (from self-playing pianos) to create unbreakable codes. It was not used at the time, but later led on to Wi-Fi and Bluetooth.

CRACKING ENIGMA

In World War II, Nazi Germany's secret messages were coded using a machine called Enigma, invented by Arthur Scherbius in 1918. The Poles were the first to crack the Enigma codes, and they taught their methods to the British, French and Americans. This helped the Allies to win the war.

TV AND VIDEO

In TV's early days, to send an image, you had to chop it up into bits, then transmit the bits to a receiver, then rearrange the image from the bits. The first efforts were mechanical – with moving parts – and the picture was fuzzy. It wasn't till better radio technology led to electronic broadcasting that crystal-clear pictures appeared on our screens.

NIPKOW DISK

Chopping up images began with an 1884 invention by German Paul Nipkow. The Nipkow disk was a spinning circle with a spiral of holes. It projected images onto a photo-electric sensor, which turned them into light/dark 'messages'. A second Nipkow disk unscrambled the messages to recreate the image.

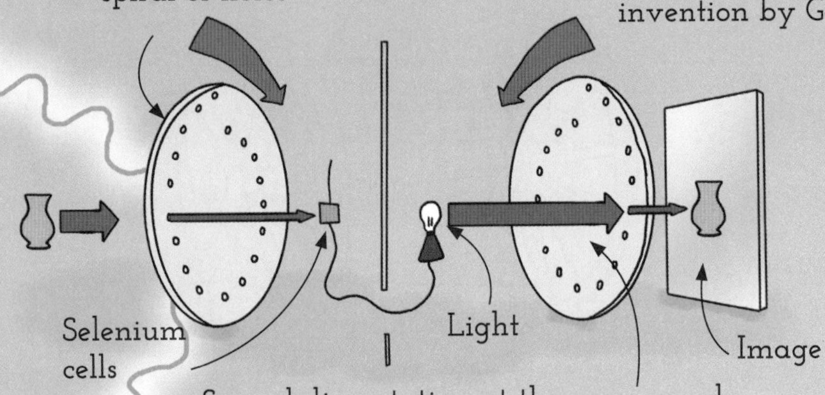

JOHN LOGIE BAIRD

The father of TV was arguably Scotsman John Logie Baird. He was so poor, and he built his early equipment from junk. In 1928 he made the first transatlantic TV broadcast, from London to New York. The next year, he began broadcasting for the new BBC television service.

JENKINS TV

In America, C. Francis Jenkins designed a Nipkow-style TV in 1923 and launched the US's first TV station in 1928. Francis also invented the waxed paper carton (for juice or milk) and an aircraft catapult.

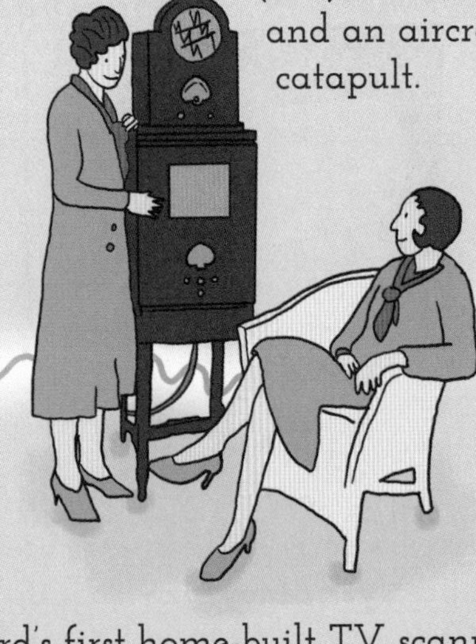

Baird's first home-built TV scanner included some Nipkow disks, an old hatbox, darning needles, bike lights and plenty of glue.

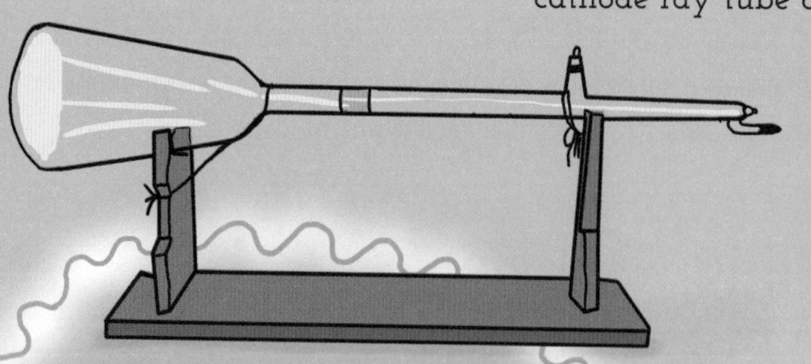

CATHODE RAY TUBE

Old TV sets – the fat, heavy ones – contained a cathode ray tube or CRT. Inside this, a beam of electrons shone at a phosphorescent (glowing) surface – the screen – to make the image. German scientist Karl Braun invented the first CRT in 1897, and Russian-born Vladimir Zworykin improved on it in the 1920s (example pictured). With the invention of the CRT, television had become electronic.

FLASH-MATIC

The world's first TV remote control was the Zenith 'Lazy-Bones' of 1950. A wire connected it to the set. Much more space-age was Zenith's 'Flash-Matic' of 1955, which shone a 'magic light' at your TV to change the channel or volume. (Unless of course your dog was sitting in the way.)

Umm, Flash, geddout the way!

I can't see a thing, Bing!

VIDEO CASSETTE RECORDER

Before the invention of the DVD in 1995, the video cassette recorder (VCR) was used to record and play back programmes. American singer Bing Crosby wanted to record his shows, so he helped the Ampex company develop the first VCR, the VR-1000, in the 1950s. It was as big as a gas cooker!

COMPUTERS

The computer is perhaps the single most important invention of our time. The first computers, invented less than a century ago, were like dinosaurs – big body, small brain – but not so today. A single modern smartphone is millions of times more powerful than all of NASA's computers from 1969. (And they managed to put men on the Moon!)

WEAVING ORIGINS

The idea for the computer program came from the French weaving industry over 250 years ago. To weave a pattern on a Jacquard loom, you fitted it with a chain of cards, each punched with a set of holes. The patterns of holes told the machine which threads to pick.

MECHANICAL COMPUTERS

Englishman Charles Babbage (1791-1871) designed the first mechanical computers: the 'Difference' and 'Analytical' engines. They were huge, with thousands of moving parts, for doing tough maths. But he was never able to complete their construction.

COMPUTER WHIZZES

The first computer programmer was Ada Lovelace (below), daughter of famous poet Lord Byron. She worked for Charles Babbage. Grace Hopper (1906-92), a rear admiral of the US Navy, wrote computer code, including the COBOL language. It's said she also invented the term 'computer bug', when a moth fell into her equipment.

DATA

The US Census Bureau collected information on American citizens, such as birth and marriage dates. To help sort the data, Herman Hollerith, a Bureau worker, invented a series of 'tabulating' machines from the 1890s onwards. These used hole-punched cards to hold the data.

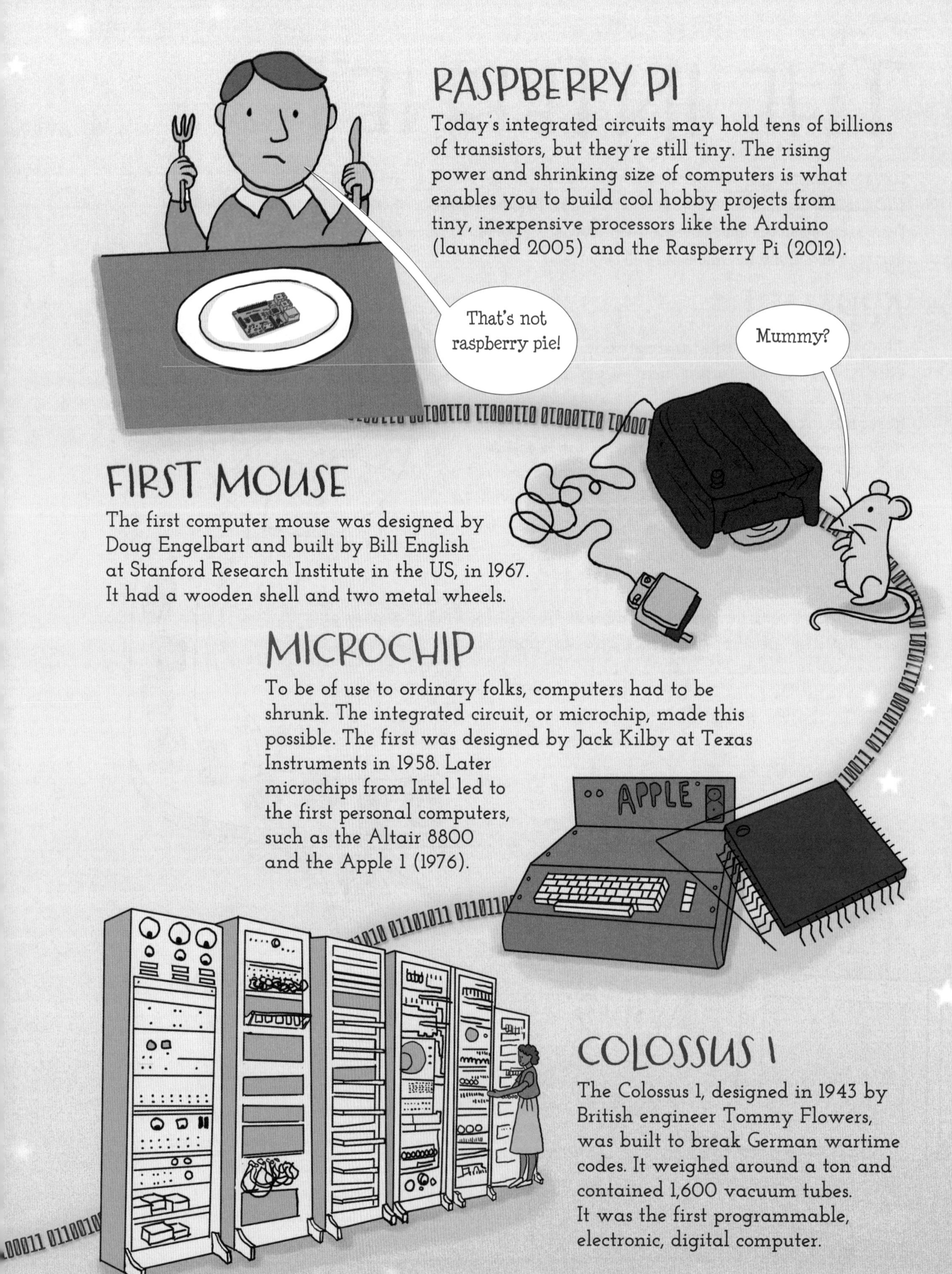

RASPBERRY PI

Today's integrated circuits may hold tens of billions of transistors, but they're still tiny. The rising power and shrinking size of computers is what enables you to build cool hobby projects from tiny, inexpensive processors like the Arduino (launched 2005) and the Raspberry Pi (2012).

FIRST MOUSE

The first computer mouse was designed by Doug Engelbart and built by Bill English at Stanford Research Institute in the US, in 1967. It had a wooden shell and two metal wheels.

MICROCHIP

To be of use to ordinary folks, computers had to be shrunk. The integrated circuit, or microchip, made this possible. The first was designed by Jack Kilby at Texas Instruments in 1958. Later microchips from Intel led to the first personal computers, such as the Altair 8800 and the Apple 1 (1976).

COLOSSUS 1

The Colossus 1, designed in 1943 by British engineer Tommy Flowers, was built to break German wartime codes. It weighed around a ton and contained 1,600 vacuum tubes. It was the first programmable, electronic, digital computer.

THE INTERNET

Just a few decades ago, the Internet was a bit like a club, open only to certain groups, such as scientists and government departments. Today, going online is as easy as picking up the phone. Every day, more than a billion people use Facebook; every minute, over 300 hours of video are uploaded to YouTube. But where did it all begin?

ARPANET

The first version of the Internet was ARPANET. The network was set up in 1969 just for DARPA: the Defense Advanced Research Projects Agency. This US government department of inventors wanted to share ideas between chosen groups. One way of sharing was by email, which was invented @ ARPANET by Ray Tomlinson in 1971.

PACKET-SWITCHING

The Internet communicates by packet-switching. This is a way of sending little electronic 'packets' of data along shared pathways. It was created in the 1960s by Leonard Kleinrock and Paul Baran in the US, and Donald Davies in the UK. In 1973, Vint Cerf and Bob Kahn in the US wrote TCP: the 'rules' for packet-switching. That year, too, ARPANET became 'the Internet', and went international.

WORLD WIDE WEB

The World Wide Web (WWW) is all the information on the Internet. It was created by English computer scientist Tim Berners-Lee. In the 1980s, Tim worked with CERN, a European nuclear research organization. He invented HTML (Hyper Text Markup Language), and CERN began using it, on what was to be the world's first web page.

INSTAGRAM

Kevin Systrom and Mike Krieger invented Instagram, the photo- and video-sharing service, in 2010. Two years later, Facebook bought Instagram for roughly $1 billion.

FACEBOOK

The social networking website, Facebook, was invented by American student Mark Zuckerberg in 2004. He first named it 'Facemash', but luckily he changed his mind. Four years later, he became the world's youngest ever billionaire.

RFID

The IoT uses RFID – radio frequency identification. Credit cards, shop security tags and microchips in pets and farm animals all use RFID to connect via electromagnetic fields. Many people helped create RFID. For instance, Russian inventor Léon Theremin used it for a bugging device for spying on the Americans, in 1945.

INTERNET OF THINGS

One day, your family's car, fridge, computers and maybe even your pet will all be hooked up to the Internet. It's called the Internet of Things (IoT), and it's already begun. The first 'thing' on the IoT was a soft-drinks machine at Carnegie Mellon University in the US, in 1982. Students checked online whether it contained chilled drinks, to save checking for real.

WACKY INVENTIONS

The brilliant English scientist Stephen Hawking once said, 'Mankind's greatest achievements have come about by talking, and its greatest failures by not talking.' So true – but talking isn't always easy. Just look at some of these crazy communication inventions.

DIAL-A-GHOST

It is reported that, in 1920, Thomas Edison, one of America's most famous inventors, proposed making a 'spirit telephone' for talking to ghosts. This story is quite probably made up – but it's a good one.

BAIRD IDEAS...

John Logie Baird came up with good inventions – like TV and thermal socks – but some bad ones too. For example:

• Glass razor blade. His idea was to make a razor that wouldn't go rusty. Trouble was, it just shattered. Ouch!

• Artificial diamonds. Diamonds are a form of carbon. Graphite (pencil 'lead') is also a form of carbon. Diamonds are made naturally underground, by high pressure and temperature. Baird figured he could make them by heating graphite. But all he made was trouble when his experiment cut the power to the city of Glasgow, leaving everyone in the dark.

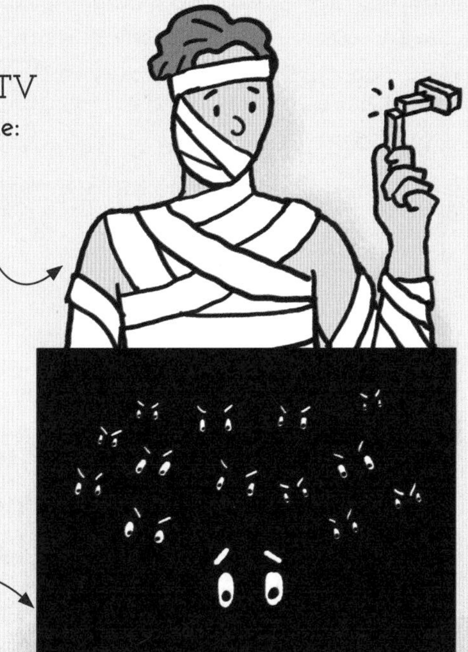

HOT TENNIS BALLS

American inventor Marvin Middlemark made millions from his Rabbit Ears TV antenna of 1953. He was less successful with a water-powered potato peeler and a plan to rejuvenate tennis balls by microwaving them.

BOW-LINGUAL

Want to chat to your dog? In 2002, a Japanese team invented a machine that listens to your dog's barking, and translates it into an emotion (such as happy or sad), which is shown on a screen. Different versions have been made for different doggy dialects around the world.

PAWSENSE

You know when the cat walks across the keyboard, turning your homework into gobbledy-gook? This piece of computer software could save you. Whenever it detects cat-like typing, it sounds an alarm (which annoys the cat and drives it away) and 'locks down' the keyboard.

CAMPILLO LOONYGRAPH

Some early versions of the telegraph – and there were many of them – were a bit crazy. In 1804, Spanish scientist Francisco Salva Campillo suggested connecting a group of people to electric wires. Each person stood for a different letter or number. Messages coming down the wires would shock each person, who would cry out their letter, spelling out the message. Eek!

DAFT PREDICTIONS

It's easy to look back through history and laugh at some of the crazy things people believed or said. At the time, of course, they didn't know any better, so it's not really their fault. But some of these quotes might still make you chuckle!

THE FUTURE OF MOVIES

'The cinema is little more than a fad.'
–Charlie Chaplin, actor and film director, 1916

THE FUTURE OF MOVIE SOUND

'Who the hell wants to hear actors talk?'
–Harry Warner, president of the Warner Bros movie company, 1925

(Today, of course, 99.999999999 per cent of movies are 'talkies'.)

THE FUTURE OF THE TELEPHONE

'The Americans have need of the telephone, but we do not. We have plenty of messenger boys.'
–William Preece, chief engineer at the British Post Office, 1879

(Today, more than 33 million families in the UK have a landline, and 93 per cent of adults own a mobile phone.)

THE FUTURE OF WAR

'The coming of the wireless era will make war impossible, because it will make war ridiculous.'
—Guglielmo Marconi, radio pioneer, 1912

(It may be ridiculous, but, sadly, wireless has only helped people wage war.)

THE FUTURE OF COMPUTERS

'There is a world market for maybe five computers.'
—Thomas Watson, head of computer company IBM, 1943

(Today, there are more than two billion personal computers around the world.)

'Computers of the future may weigh no more than 1.5 tons.'
—Popular Mechanics magazine, 1949

(Well, at least they were right about that – thank goodness!)

THE FUTURE OF THE INTERNET

'I predict the Internet will soon go spectacularly supernova and in 1996 catastrophically collapse.'
—Robert Metcalfe, founder of computer network company 3Com, 1995

(Bob promised to eat his words if he was wrong – which, as we now know, he was! Two years later, he used a kitchen blender to liquefy a printout of his prediction, and then drank it.)

INDEX

The Author

British-born Matt Turner graduated from Loughborough College of Art in the 1980s, since which he has worked as a picture researcher, editor and writer. He has authored books on diverse topics including natural history, earth sciences and railways, as well as hundreds of articles for encyclopedias and partworks, covering everything from elephants to abstract art. He and his family currently live near Auckland, New Zealand, where he volunteers for the local Coastguard unit and dabbles in art and craft.

The Illustrator

Sarah Conner lives in the lovely English countryside, in a cute cottage with her dogs and a cat. She spends her days sketching and doodling the world around her. She has always been inspired by nature and it influences much of her work. Sarah formerly used pens and paint for her illustration, but in recent years has transferred her styles to the computer as it better suits today's industry. However, she still likes to get her watercolours out from time to time, and paint the flowers in her garden!

INCREDIBLE INVENTIONS
LET'S COMMUNICATE

Written by Matt Turner
Illustrated by Sarah Conner

Yay! Only 1,285 pages still to print...

外语教学与研究出版社
FOREIGN LANGUAGE TEACHING AND RESEARCH PRESS
北京 BEIJING

奇思妙想发明史

五花八门的交通工具

马特·特纳（英）著
萨拉·康纳（英）绘
周渝毅 译

外语教学与研究出版社
FOREIGN LANGUAGE TEACHING AND RESEARCH PRESS
北京 BEIJING

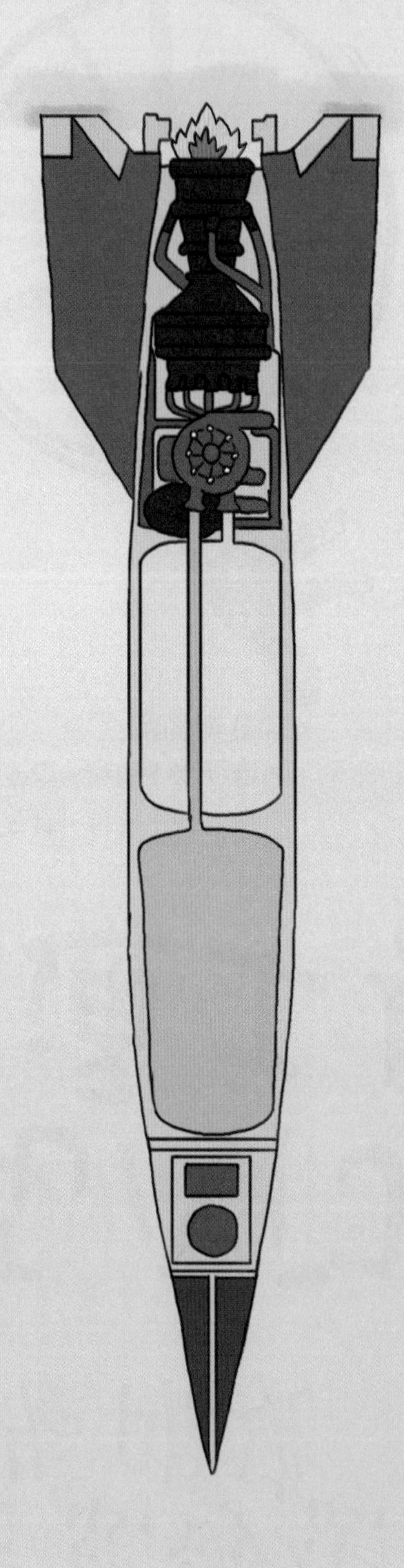

让我们一起来回顾一下道路交通工具（从双轮马车到大众甲壳虫汽车）以及海上交通工具（从独木舟到航空母舰）的发展进程吧。你还可以了解铁路技术的革命，以及人类征服蓝天的历史——从中国人发明的简易风筝到喷气式飞机。后来人类又听到了太空的召唤……在这本书里，你会发现：

- 卡尔·本茨造的第一辆“汽车”只有三个车轮
- 一张三角帆带来了航海时代的巨变
- 蒸汽机车诞生于200多年前，那时候还很慢，时速只有3.9千米
- 现代太空火箭的创意最早出现于1903年

还有更多的发现等着你。不要错过那些天马行空且从未实现的创意（比如核动力汽车和会飞的潜艇），也不要错过那些成功的发明，比如惊险刺激的喷气背包和翼服……

2015年，日本磁悬浮列车在一次测试中创下了时速605千米的纪录。

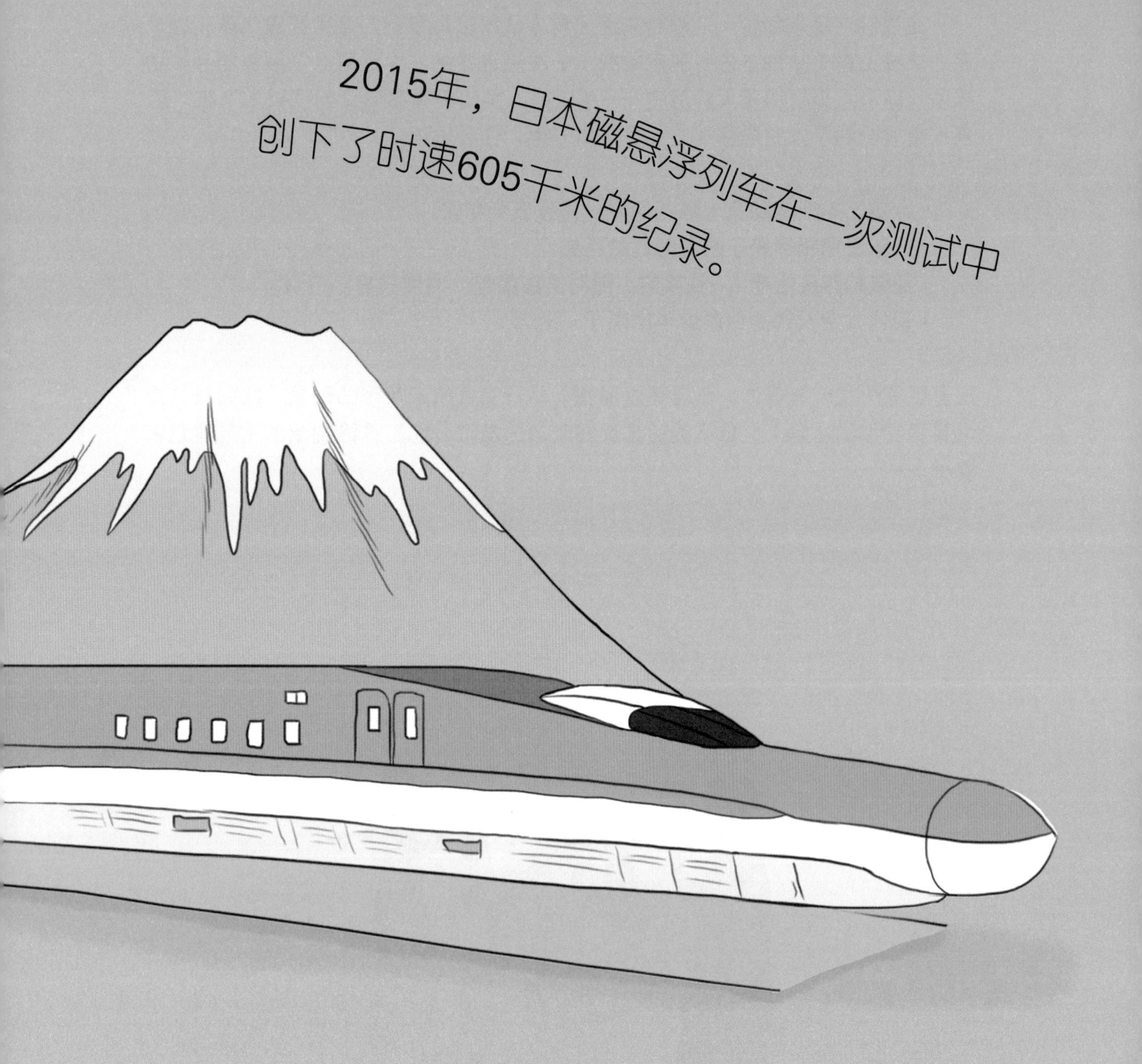

目录

导语：我们出发了！

人类是天生的行者，但数万年来我们的祖先去哪儿都是依靠双脚。他们没有别的交通工具，至少在驯化野生动物前是这样。比如早期人类大约在6,000年前驯化了马。许多年以后，人们发明了车轮——由此产生了手推车和马车。人们还利用风力驱动帆船来横渡海洋。

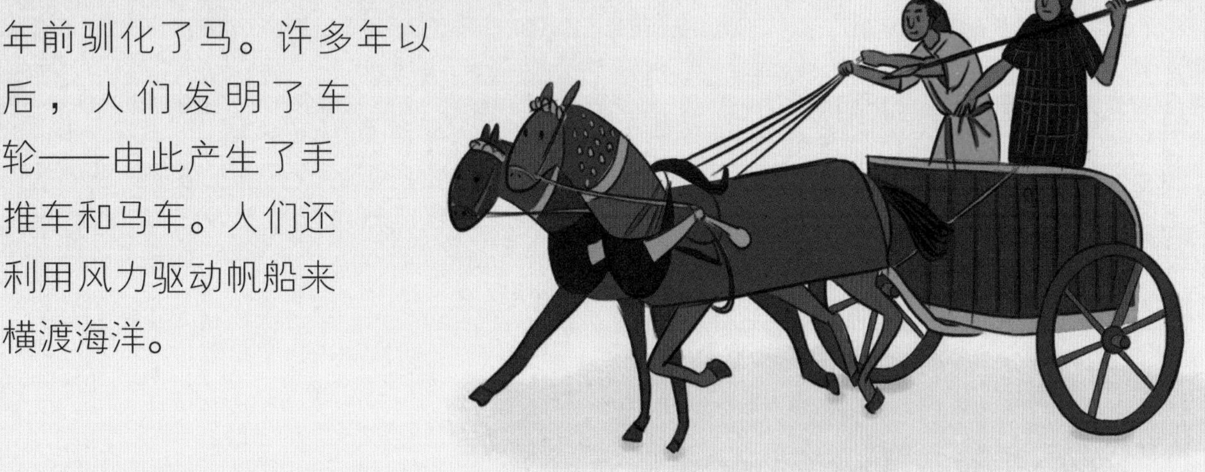

辐条轮，公元前2,000年

最近200年来，随着蒸汽发动机和内燃机的发展，交通工具史才发生了翻天覆地的变化。轮船、火车、汽车和最晚发明的飞机都用蒸汽发动机和内燃机提供动力。这些交通工具为我们带来了便利——速度和舒适性都提升了——但也带来了问题，比如污染。顺便说一下，你知道吗，最初汽车被视为可以取代马车的“清洁型”交通工具。因为城市街道上成吨的马粪会招来苍蝇，传播疾病。当然，现在汽车尾气很呛人，我们正在寻找替代能源，比如电。（英国养鸡场场主哈罗德·贝特在1971年发明了贝特汽车，这种汽车貌似更有利于健康且更环保，因为它以动物粪便为动力！）

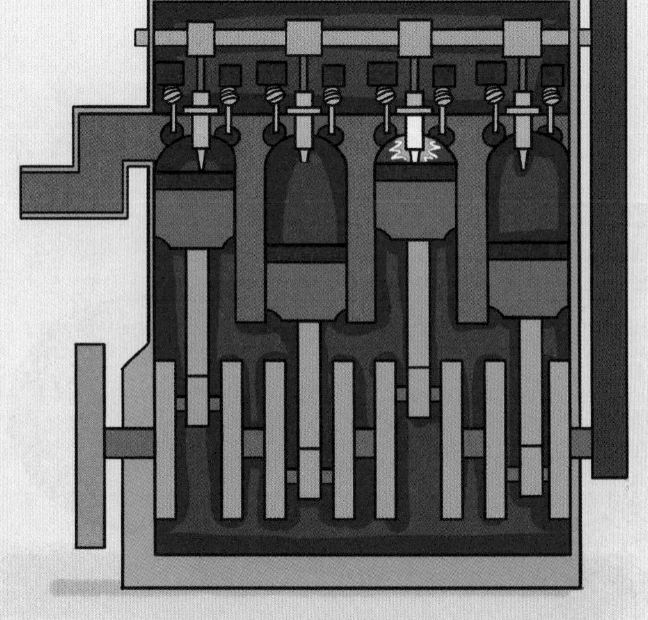

内燃机，19世纪50年代

世界上最早的航空先驱从鸟类身上找到灵感，但他们发明的扑翼飞行器很失败。直到莱特兄弟在1903年完成历史性飞行，人类才揭开了载人航空飞行的奥秘。后来人们在战争中使用火箭，招致毁灭性后果后，开始了对人类最后的边疆——太空——的和平探索。

独木舟，公元前6,000年（或更早）

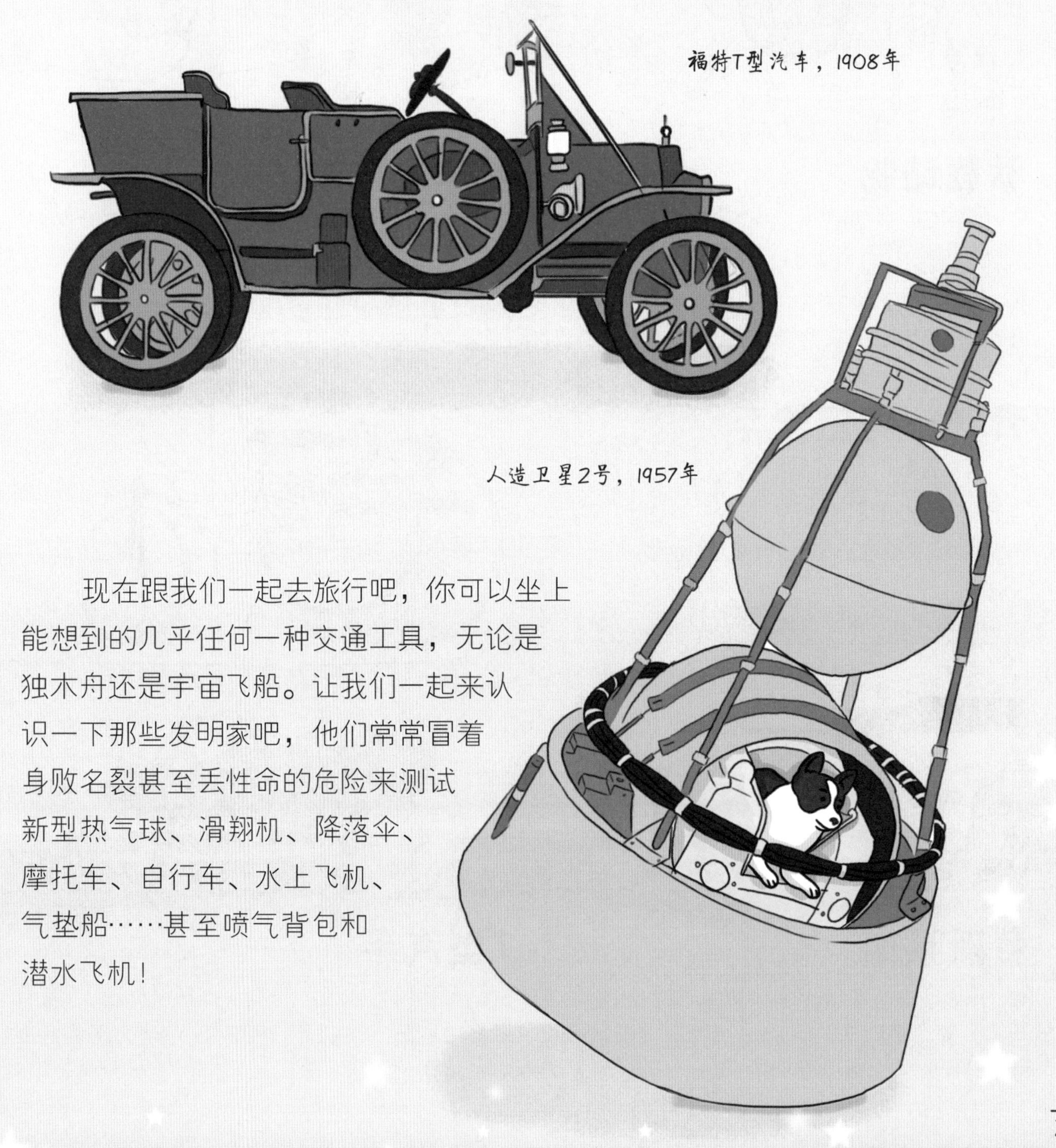

福特T型汽车，1908年

人造卫星2号，1957年

现在跟我们一起去旅行吧，你可以坐上能想到的几乎任何一种交通工具，无论是独木舟还是宇宙飞船。让我们一起来认识一下那些发明家吧，他们常常冒着身败名裂甚至丢性命的危险来测试新型热气球、滑翔机、降落伞、摩托车、自行车、水上飞机、气垫船……甚至喷气背包和潜水飞机！

车轮和车的发明

车轮的发明引发了交通工具和科技的革命性巨变，但在人类历史长河中，车轮的发明来得相当晚——大约公元前3,000年才出现，大大晚于人类发明矛、笛子和陶器的时间。事实上，人们发明车轮是从陶工旋盘中获得了启发。车轮这么晚才问世，很可能是因为自然界没有轮子，轮子完全是人类的发明。

驮载动物

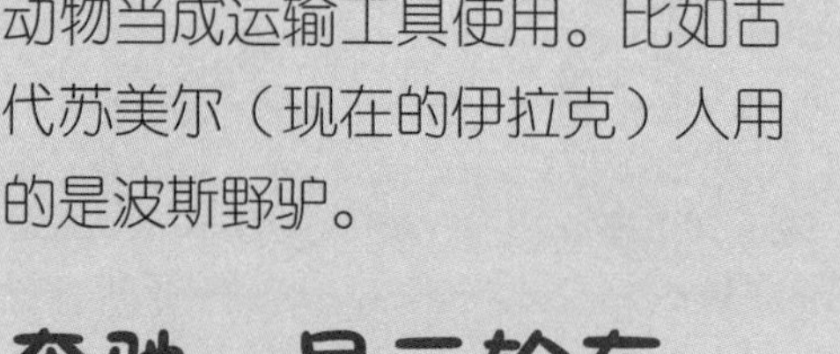

在车轮出现前，人们将驮载动物当成运输工具使用。比如古代苏美尔（现在的伊拉克）人用的是波斯野驴。

最早的车轮

最早的车轮由切割后的实心木板拼接而成。上图所示的牛车于公元前2,500年左右在欧洲北部出现，这种车是用家养的牛拉动的。

奔驰一号三轮车

1885年，德国发明家卡尔·本茨（后来以梅赛德斯—奔驰闻名）制造了奔驰一号三轮车。他的妻子贝尔塔在开这辆车去探望年迈母亲的路上，发明了刹车片。

劳斯莱斯

1907年，劳斯莱斯40/50银色幻影问世。一家汽车杂志称它为“世界上顶级的汽车”，它也是当时最昂贵的汽车。

T型汽车

1908年，美国人亨利·福特设计出了T型汽车。1913年，他发明了这种汽车的生产线：汽车严格按照顺序生产——每分钟可产出三辆！

辐条轮

4,000年前，安德罗诺沃人驾驶双轮战车驰骋在欧亚平原上。据说是他们发明了辐条，辐条使车轮更轻、更大，速度也更快。

驿马车

驿马车是在匈牙利的科奇发明的。17世纪的欧洲，人们在长途旅行时一般会乘坐驿马车。早期的驿马车坐起来很颠簸，但是1800年弹簧减震装置和刹车的发明提升了驿马车的速度和舒适性。

蒸汽发动机

蒸汽发动机从18世纪80年代起被用于船只，后来被用于机车，也开始为道路车辆提供动力。顺从号是一辆蒸汽公共汽车，是法国人阿梅代·博莱在1875年制造的。

马车

19世纪50年代晚期，比利时人艾蒂安·勒努瓦发明了以煤气为动力的发动机。后来他把这个发动机装在马车上，安装后马车的最高时速为三千米——比步行速度还慢。

Volkswagen的意思是“国民的汽车”。截至2003年，售出的“甲壳虫”汽车已超过2,100万辆。

嗯，给它起个什么名字好呢？

“甲壳虫”汽车

德国的保时捷公司生产豪华跑车。但是早在20世纪30年代，费迪南德·波尔舍还设计了大家都能买得起的“大众甲壳虫”汽车——因其外形酷似虫子而得名。

内燃机

内燃机以汽油、柴油或煤气为燃料。燃料在汽缸中爆炸后释放出能量，从而驱动车轮运转。内燃机已有150多年的历史，自问世之日起经历了很多改进。

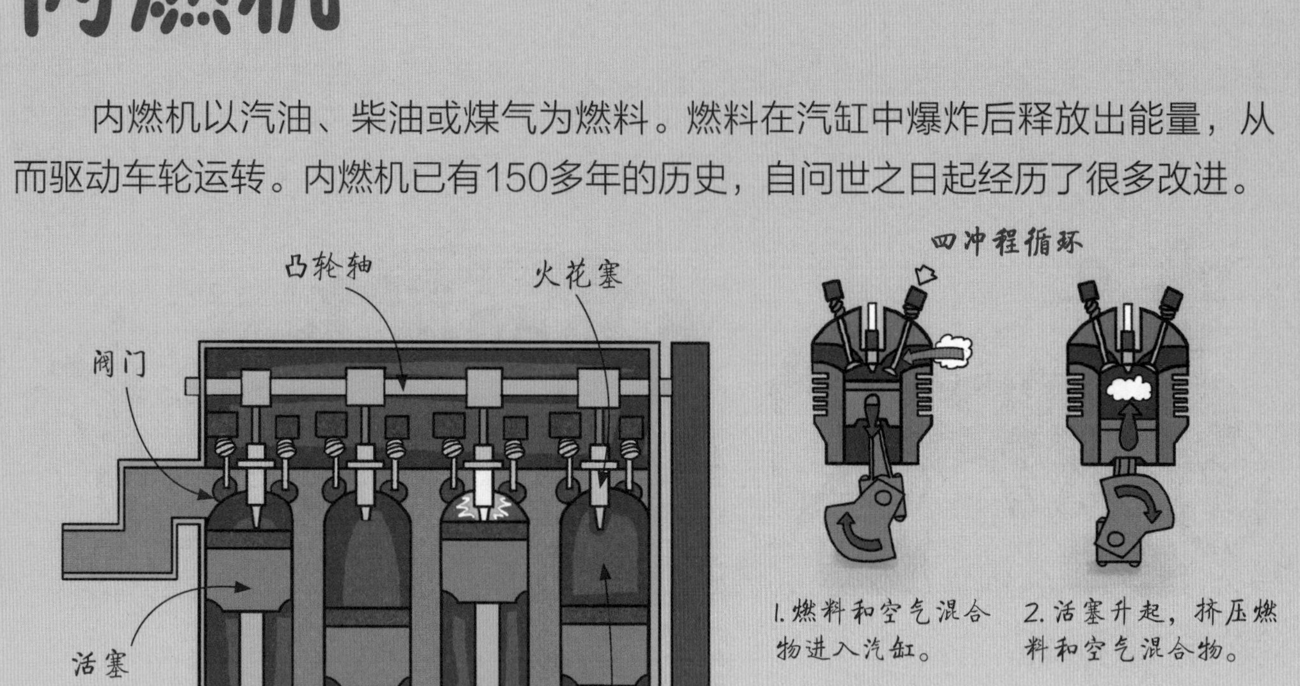

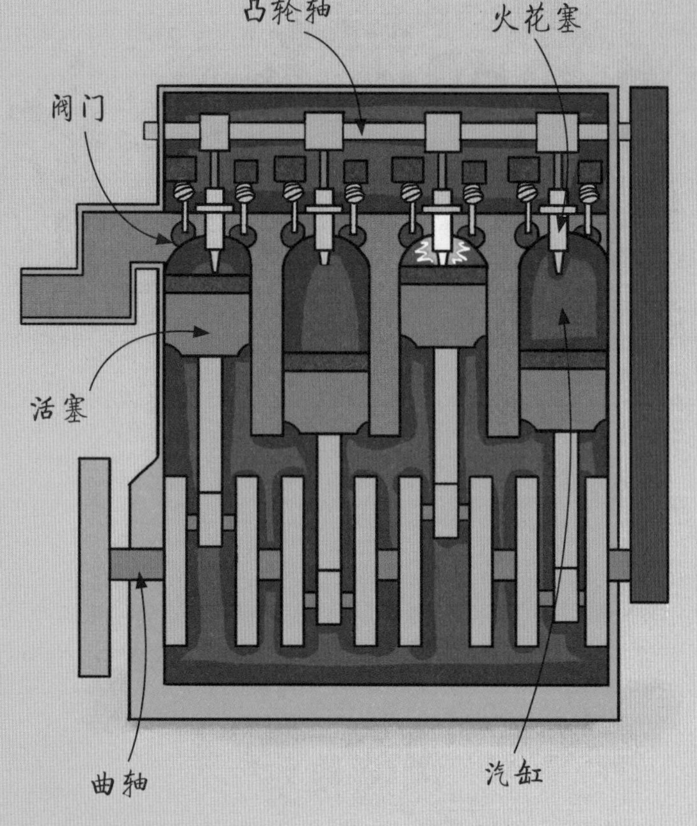

1.燃料和空气混合物进入汽缸。

2.活塞升起，挤压燃料和空气混合物。

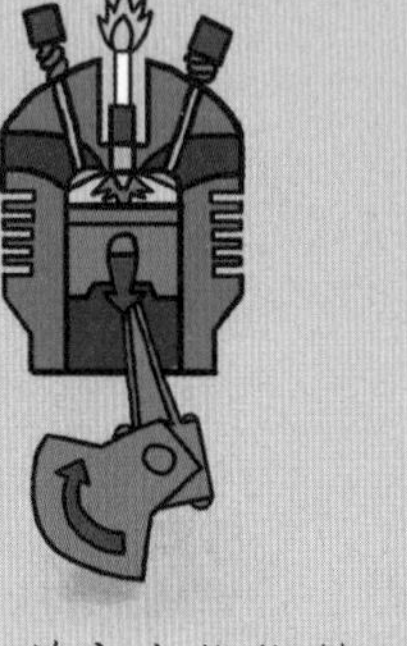

3.火花塞点燃燃料，迫使汽缸下行。

4.活塞回位，把废气（燃烧后的气体）排出。

化油器

燃料在化油器（如右图所示，其工作原理有点像老式的香水喷雾器）里被汽化。然后燃料雾气进入汽缸顶部，并在此由火花塞点燃。世界上第一台高效化油器发明于19世纪80年代。

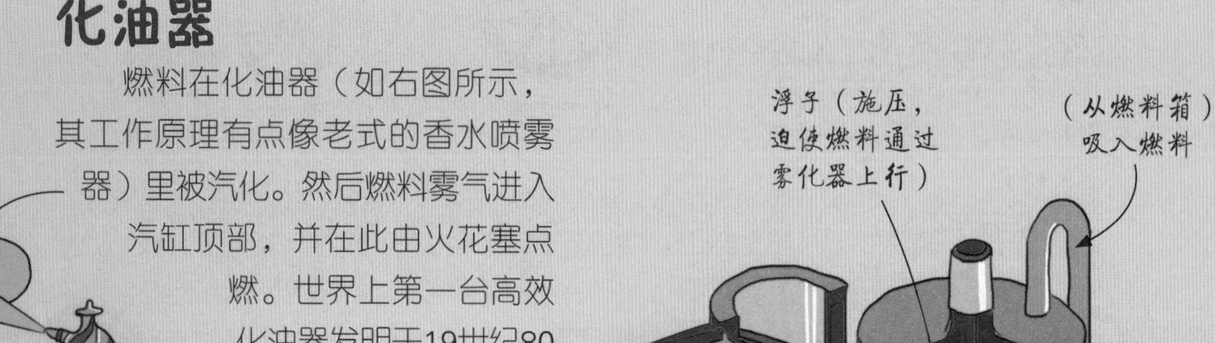

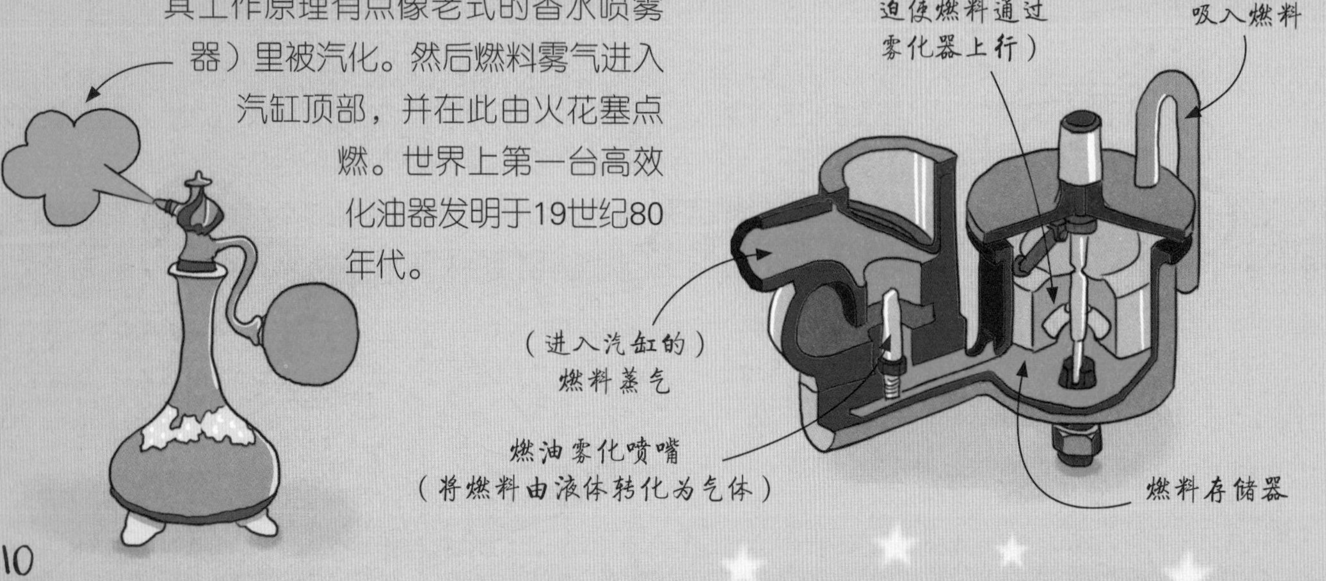

煤气发动机

还记得艾蒂安·勒努瓦和他的马车吗？右图是他在19世纪50年代发明的发动机，燃料是当时用于路灯照明的煤气。但差不多就在那个时候，人们开始把原油转化为石蜡、煤油和汽油。汽油成为更受欢迎的发动机燃料。

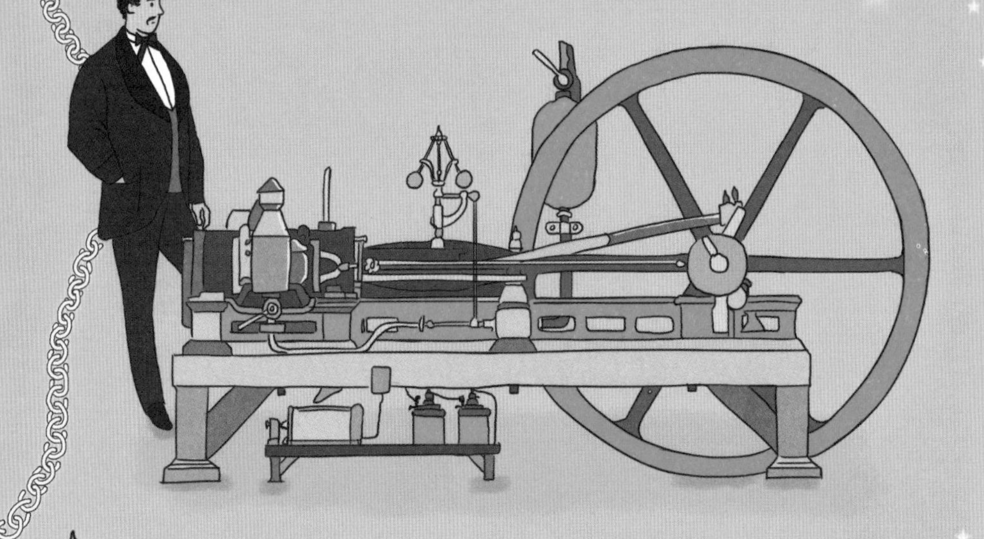

我要叫它“狄塞耳”！

柴油发动机

1892到1897年间，德国人鲁道夫·狄塞耳发明并改进了以自己的名字命名的柴油发动机。他制造的第一台柴油发动机的效率是蒸汽发动机的七倍——使用的燃料更少，但产生的动力更大。

V8发动机有八个汽缸。这些发动机分成两排，呈V字形，每排各四个。

哇……哦……好快啊！

V8发动机

1902年，法国人莱昂·勒瓦瓦瑟尔发明了V8发动机，并将其命名为“安托瓦妮特”（以其赞助商女儿的名字命名）。V8发动机主要用在飞机里。右图这款达拉克赛车生产于1905年，配备了22.5升的V8发动机，打破了当时陆地行驶的最快纪录！

从自行车到摩托车

你是不是很难想象一个没有自行车的世界会怎样？但自行车的历史大约只有200年。最早出现的“震骨头”是用木头和铁做的，坐上去硬邦邦的。摩托车大约出现在150年前，刚问世时也很原始。

德莱斯老式自行车

1817年发明的木制德莱斯老式自行车是以其德国发明者卡尔·冯·德莱斯的名字命名的。它没有踏板，德莱斯称之为“奔跑的机器”。

早期脚踏车

19世纪60年代，皮埃尔·拉勒芒和皮埃尔·米肖设计了法式米肖自行车，也称脚踏车。这种车没有轮胎，但座板装有弹簧。

摩托车

1884年，英格兰人爱德华·巴特勒发明了上图这种三轮摩托车。但它最高时速为16千米，不能进入建筑物密集的地区；当时市区的限速为时速3千米。

骑行摩托车

1885年，戈特利布·戴姆勒和威廉·迈巴赫造出了骑行摩托车，这是世界上第一辆真正意义上的摩托车。但是它的座板着火了。哎哟！

早期的摩托车

1894年，德国制造的希尔德布兰德&沃尔夫穆勒两轮摩托车是世界上最早的性能可靠的摩托车之一。

凯旋摩托车

1915年，英格兰设计出凯旋摩托车H型，在第一次世界大战中供信使使用。

前轮大后轮小的自行车

19世纪70年代，常见自行车也叫“前轮大后轮小的自行车”，因为它的两个车轮看起来就像两枚硬币，一个大，一个小。

安全自行车

前轮大后轮小的自行车很难骑，骑车人经常会摔跤，后来这种自行车就被“安全自行车”取代了。下图是英格兰人约翰·斯塔利在1885年制造的罗孚安全自行车。

米肖摩托车

1867年，米肖蒸汽动力摩托车问世，这种车基本上就是一辆安装了蒸汽发动机的米肖脚踏车。

头盔

头盔是能救命的帽子。1935年，英国著名军人T. E.劳伦斯在一场摩托车车祸中丧命。之后，医生们开始呼吁使用防撞头盔。在美国，赛车手罗里·里克特的一个朋友死于车祸，在这一事故的驱动下，他于20世纪50年代创立了贝尔头盔品牌。

黄蜂牌小型摩托车

1946年，意大利制造的黄蜂牌MP-6型“机动自行车”是一种人人都能骑的小型摩托车。

本田

1969年制造的本田CB750型是日本最早的一批现代快速摩托车。

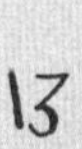

最早的船

人类自从首次划着砍伐的原木渡过小河后，就一直在不断改进船只的设计。大约5,000年前，美索不达米亚人首次使用了帆。大约1,800年前发明的大三角帆使船员能在任何风向下出海。随着航运的发展，海上强国开始出现，海上贸易日渐繁荣。

独木舟

至少在8,000年前，独木舟就作为一种简单的水上交通工具在全世界范围内被使用。首先选一根又长又直的树干，然后把中间部分的木头凿掉（或烧掉）。现在，拿起你的桨开始划吧……

埃及驳船

上图这艘埃及驳船建于4,500年前，很可能就是它把胡夫法老的遗体运到了位于吉萨的墓地。在那里，这艘驳船被密封在一座金字塔里，供已故的法老来世使用。

双排桨快艇

大约在公元前800年至公元前700年间，腓尼基人发明了双排桨快艇。顾名思义，这种船每边各有两排桨。人们将双排桨快艇用作商船，也用作战船。战船上有一个尖尖的鸟嘴似的撞角，用来撞击敌船。

航空母舰

现代航空母舰都是巨无霸，可以让飞机从甲板上起飞。一些航空母舰最长可达333米，最多能搭载90架飞机。下图这艘航空母舰被涂上了迷彩色。

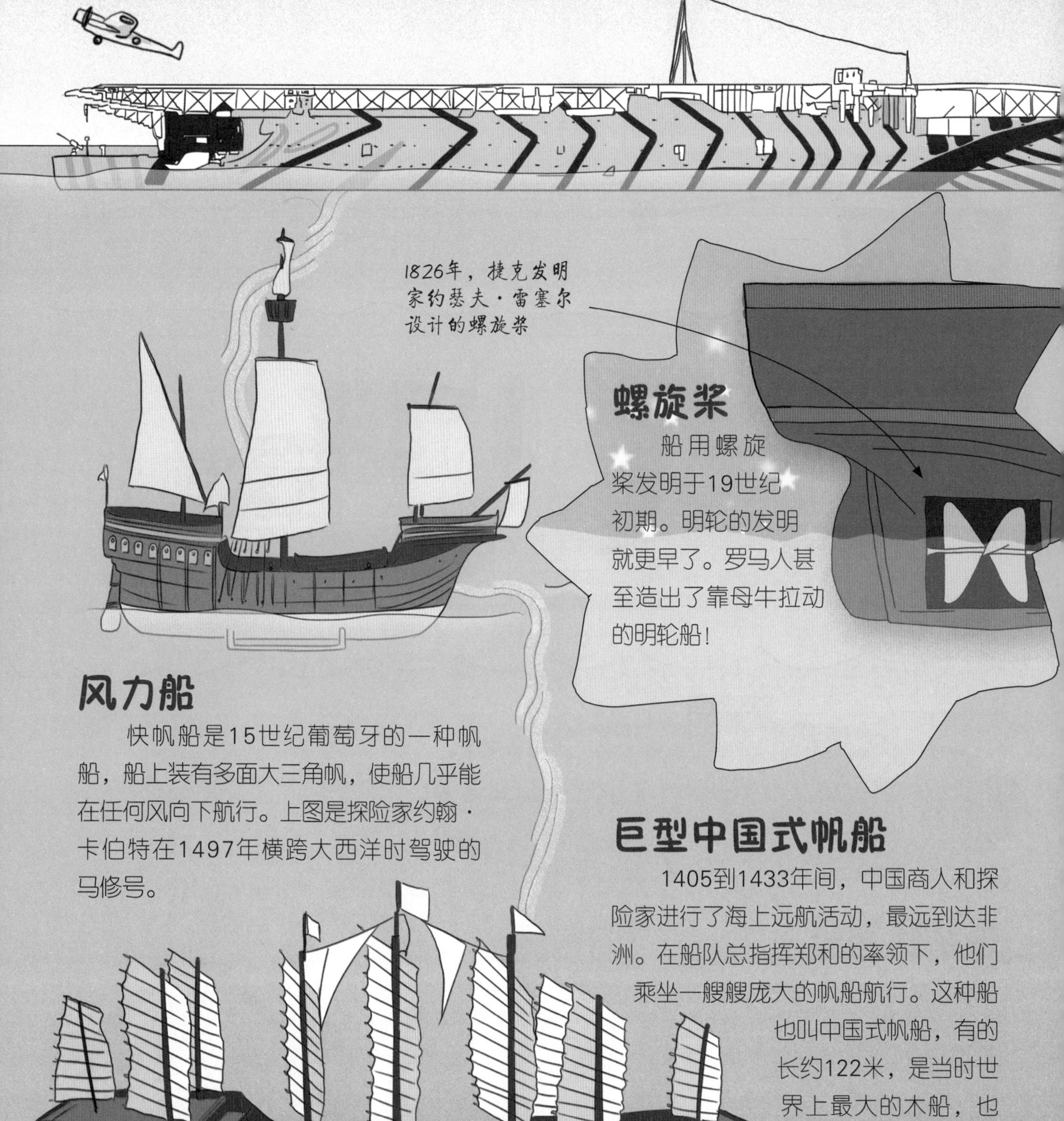

1826年，捷克发明家约瑟夫·雷塞尔设计的螺旋桨

螺旋桨

船用螺旋桨发明于19世纪初期。明轮的发明就更早了。罗马人甚至造出了靠母牛拉动的明轮船！

风力船

快帆船是15世纪葡萄牙的一种帆船，船上装有多面大三角帆，使船几乎能在任何风向下航行。上图是探险家约翰·卡伯特在1497年横跨大西洋时驾驶的马修号。

巨型中国式帆船

1405到1433年间，中国商人和探险家进行了海上远航活动，最远到达非洲。在船队总指挥郑和的率领下，他们乘坐一艘艘庞大的帆船航行。这种船也叫中国式帆船，有的长约122米，是当时世界上最大的木船，也常被叫做“宝船”。

水上奇迹

如果你能在水下行驶，为什么还要在水上航行呢？潜艇的历史可以追溯到400年前——但是你可能不愿意坐古代会漏水的破船！或许气垫船才能真正让你的船浮起来？

舷外发动机

左图是一款1902年法国生产的舷外发动机，但是装上这种发动机以后很难控制船的方向——就像在用一个巨大的打蛋器！1907年，挪威出生的奥勒·埃文鲁德制造了第一台真正意义上的舷外发动机。

气垫船

1959年，克里斯托弗·科克雷尔驾驶着他的第一艘标准气垫船SR.N1号横渡了英吉利海峡。科克雷尔的灵感来自于19世纪70年代有关气垫船的研究。

带桨的潜艇

科尼利厄斯·德雷贝尔是一个很聪明的荷兰人，他发明了许多东西，其中包括带桨的潜艇。1624年，他向英格兰国王詹姆士一世展示了自己最新式的潜艇，并且还用它带国王游览了泰晤士河！

海龟号

1775年，美国人戴维·布什内尔设计了海龟号——一种只能容纳一人的潜艇。他想在独立战争期间用这种潜艇把炸药偷偷放在英军的船只上。

地效飞机

地效飞机可以在离水面几米的上空飞行，底部有一个密封的空气垫。下图这架巨大的“花尾鸽”飞机建造于20世纪60年代的俄罗斯，长度超过90米。但它的平衡性能不好，所以这种设计理念并没有真正推广开来。

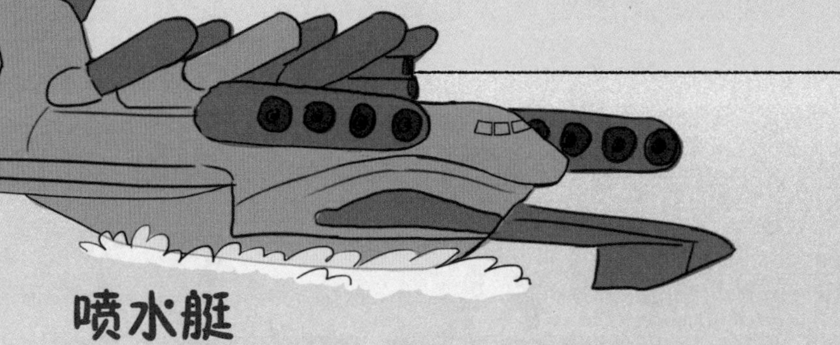

喷水艇

20世纪50年代，新西兰人比尔·汉密尔顿发明了喷水艇。喷水艇的发动机从底部吸水，然后用泵把水从后部排出，推动喷水艇前行。

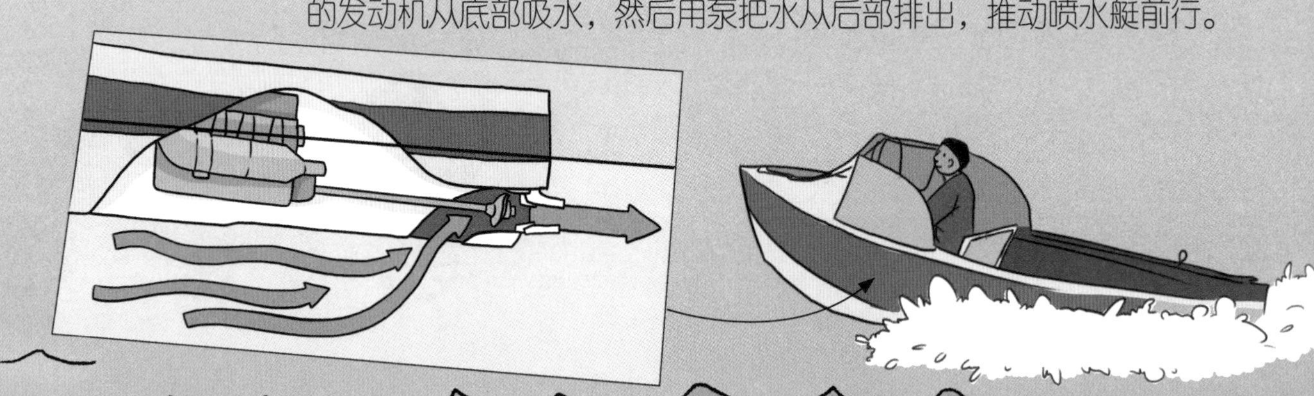

霍兰1号潜艇

下图是英国皇家海军的霍兰1号潜艇，是爱尔兰工程师约翰·霍兰在1901年设计的。1913年，霍兰1号潜艇在被拖行过程中于英国近海沉没失踪了。1983年，人们找到这艘潜艇并进行修复，后将其存放于博物馆供人们参观。

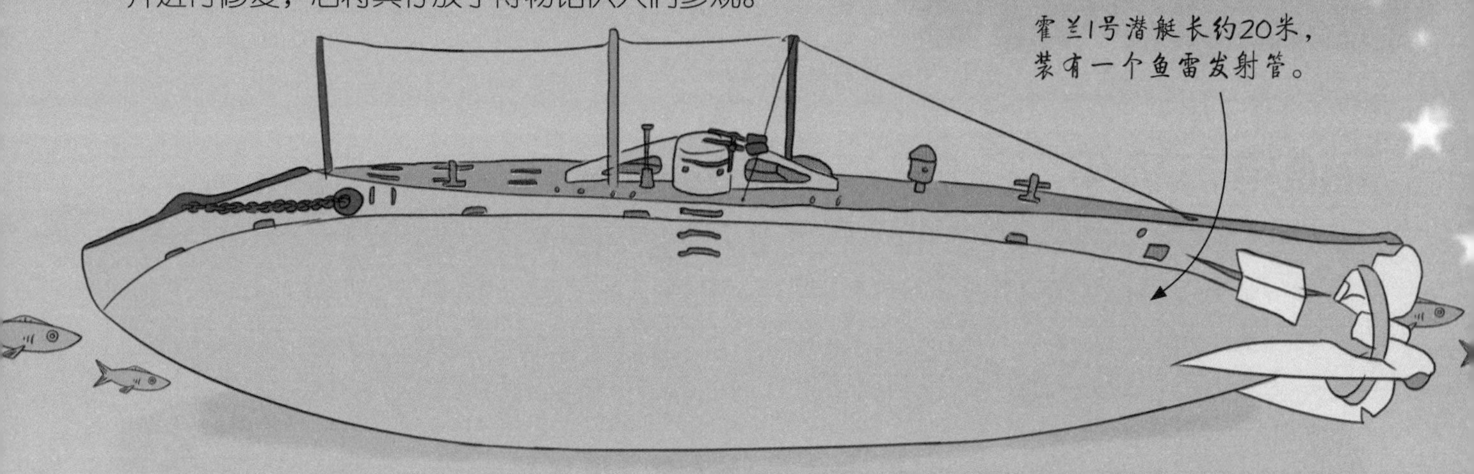

霍兰1号潜艇长约20米，
装有一个鱼雷发射管。

从蒸汽机车到磁悬浮列车

19世纪发明的动力强劲的高压蒸汽发动机很快就被应用于铁路系统，不过最早的蒸汽机车还是很慢！自那以后，柴油机车、电力机车甚至“磁悬浮”列车相继出现，带领铁路发展进入了现代化阶段，让我们享受到快捷、平稳的大众交通。

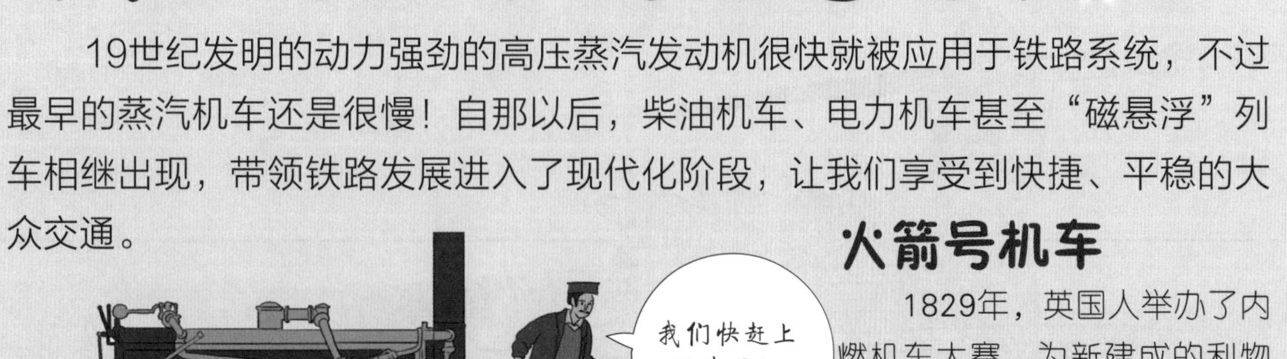

火箭号机车

1829年，英国人举办了内燃机车大赛，为新建成的利物浦—曼彻斯特铁路线选择火车型号。最终的获胜者是火箭号，由乔治和罗伯特·斯蒂芬森设计制造。这辆机车在当时算是非常现代化了。

蒸汽机车

英格兰人理查德·特里维西克设计制造了高压蒸汽发动机。1804年，他在第一条蒸汽机车铁路线上进行试验：机车以平均3.9千米的时速行驶了将近16千米。

子弹头火车

今天，日本的新干线“子弹头火车”比其时速320千米的限速快得多。在中国，磁悬浮列车漂浮在磁场上，最高时速可达430千米。

20世纪20年代由通用电气公司制造的美国柴油电力机车

柴油电力机车

柴油电力机车使用柴油发动机给发电机提供动力，发电机驱动电动机运转，从而带动火车行驶。柴油电力机车取代了市区冒烟的老旧蒸汽火车后，人们晾晒的衣服就不会被熏黑了！

电气火车

1879年，德国发明家维尔纳·冯·西门子建造了下图这种小型电气火车，这也是世界第一辆电气火车。（1880年，他设计制造了第一部电梯。1881年，他在柏林设计制造了第一辆有轨电车。）

唉，有一点小啊。

哎呀，天哪！这里边太呛人了！

大都会线

世界上第一条地铁线就是大都会线，该线路1863年在伦敦建成通车。当时大都会线上使用的是蒸汽机车，烟雾很大（引起人们咳嗽），所以后来用电气火车替代了蒸汽机车。

哐当哐当开过来！

野鸭号蒸汽机车

1938年，奈杰尔·格雷斯利爵士设计了野鸭号蒸汽机车，这列英国机车的时速刚好超过202千米，创下了当时蒸汽机车时速的最高世界纪录，这个纪录现在还没被打破。野鸭号漂亮又光滑的流线型设计大大降低了风阻。

最早的飞行实验

你看到鸟儿时是不是也想飞呢？你不是第一个有这种想法的人。人们模仿鸟儿飞上天的做法最早可以追溯到古代中国。但是随着200多年前热气球的发展，以及很久以后受控动力飞行的出现，人类的飞天梦想才变成现实。

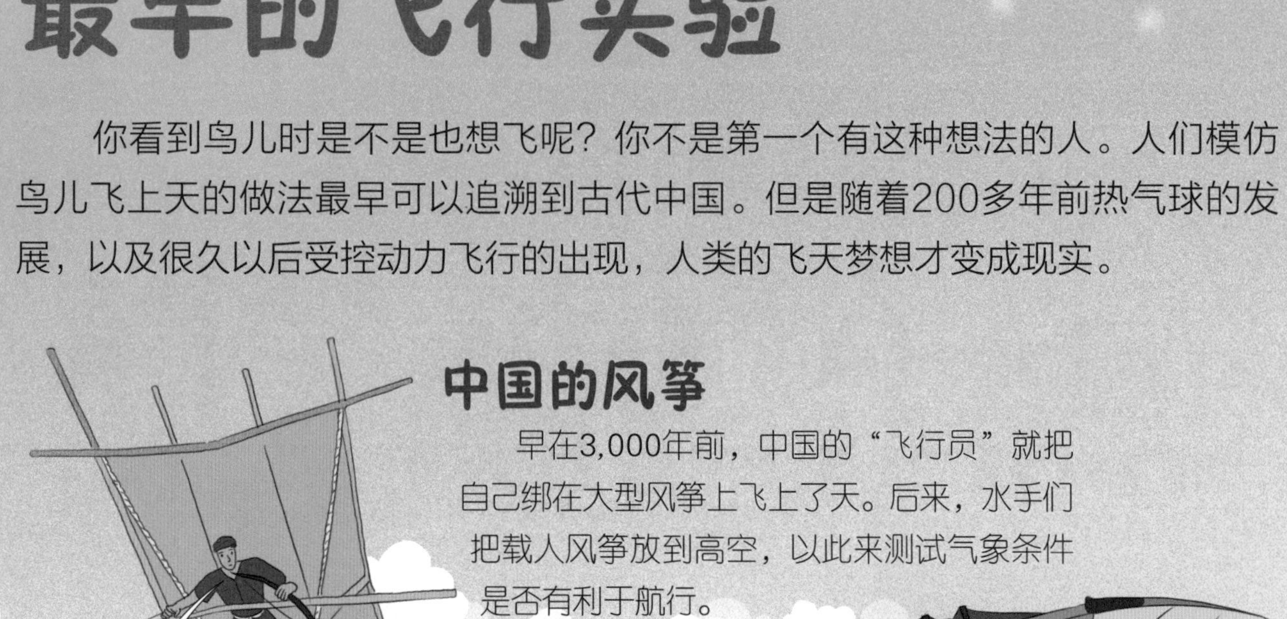

中国的风筝

早在3,000年前，中国的“飞行员”就把自己绑在大型风筝上飞上了天。后来，水手们把载人风筝放到高空，以此来测试气象条件是否有利于航行。

莱奥纳多的机器

莱奥纳多·达·芬奇（1452—1519）是意大利的天才艺术家，他受蝙蝠和鸟类的启发，发明了一种会飞行的机器，虽然实际上他并没有把它造出来。（他还发明了直升机——至少画在了纸上！）

水上飞机

水上飞机，顾名思义，就是既能从水上起飞又能在水上降落的飞机。这种飞机的历史可以追溯到1905年，当时法国飞行先驱加布里埃尔·瓦赞驾驶着一架水上滑翔机飞过塞纳河，拖拽它的是一艘快艇。

热气球

1783年9月19日的法国，蒙戈尔菲耶兄弟把一只热气球放飞上天，热气球上有一头羊、一只鸭子和一只公鸡。这些动物都安然无恙地返回了地面。当年晚些时候，蒙戈尔菲耶兄弟进行了世界上首次热气球载人飞行。

气体燃烧器

蒙戈尔菲耶兄弟在气球下面点火让气球飞上了天。1955年，美国人埃德·约斯特发明了球载气体燃烧器。1978年，约斯特的双鹰 II 号成为第一只飞越大西洋的热气球。

滑翔机

1853年，英国发明家乔治·凯莱让一名仆人驾驶一架滑翔机飞上了天。（这名仆人很幸运，毫发无损地回到了地面。）

飞艇

右图这艘巨型飞艇于1900年7月进行了首次飞行。它是以其发明者德国人费迪南德·冯·齐柏林伯爵的名字命名的。齐柏林伯爵的第一艘飞艇LZ1号长达128米，几乎是现代巨型喷气式客机长度的三倍。

飞机的发明

下次你坐飞机时，坐的很可能是舒适的喷气式客机，飞机上提供餐饮，可能还可以看电影。但是仅仅100年前，飞机还是一种易损的飞行器，是用“棍子、绳子和布料”搭成的，人们几乎是凭着希望和勇气让它们飞上天的。

喷气式飞机

第一批喷气式飞机也是在第二次世界大战中出现的，发明者是德国人汉斯·冯·奥海恩和英国人弗兰克·惠特尔。1939年，一架装有冯·奥海恩发动机的亨克尔He178喷气机进行了世界上第一次喷气式飞行。

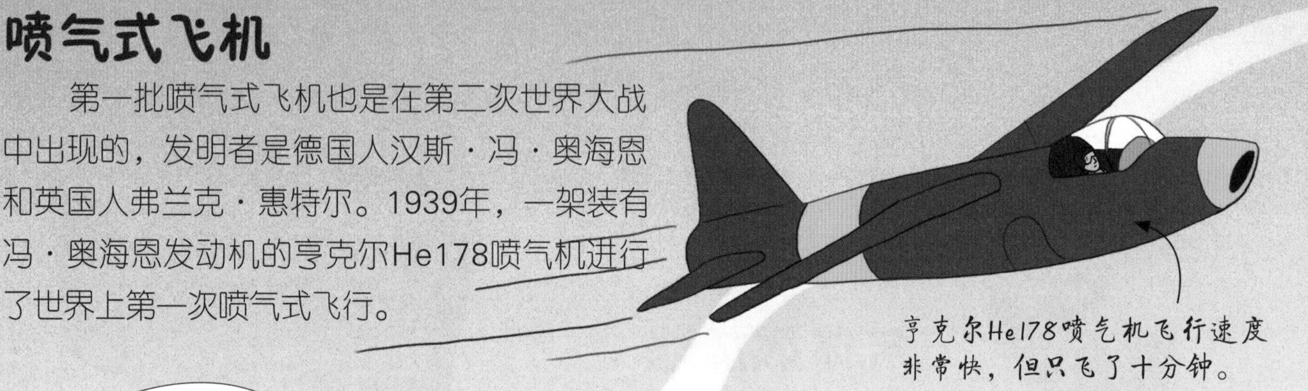

亨克尔He178喷气机飞行速度非常快，但只飞了十分钟。

喷火式战斗机

1936年，雷吉·米切尔设计出英国超级马林喷火式战斗机。它的两侧机翼都可以装载机枪，是二战期间非常有名的战斗机。最初米切尔想给它起名为“泼妇”或“圣甲虫”，他不喜欢“喷火”这个名字。当时制造了20,000多架喷火式战斗机，现在仍有大约50架在服役。

喷火式战斗机名副其实。哦，比“泼妇”这个名字好。
EBOB
H8882

飞起来了！

莱特兄弟

1903年，威尔伯·莱特和奥维尔·莱特在北卡罗来纳州的基蒂霍克进行了世界上第一次动力飞行。他们最初建造的是滑翔机，但是他们的第一架动力飞机“飞行家”1号使用了12马力的发动机。

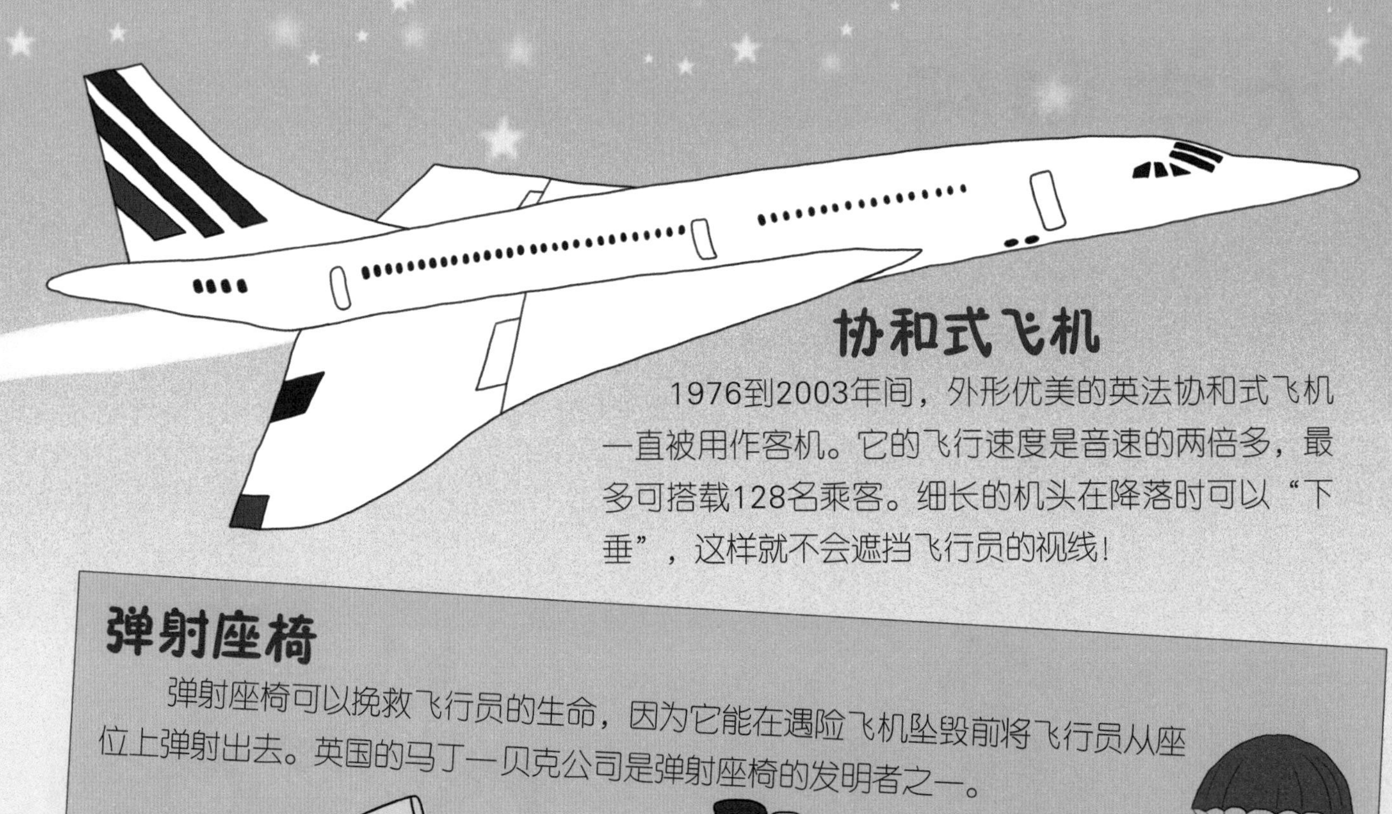

协和式飞机

1976到2003年间，外形优美的英法协和式飞机一直被用作客机。它的飞行速度是音速的两倍多，最多可搭载128名乘客。细长的机头在降落时可以“下垂”，这样就不会遮挡飞行员的视线！

弹射座椅

弹射座椅可以挽救飞行员的生命，因为它能在遇险飞机坠毁前将飞行员从座位上弹射出去。英国的马丁—贝克公司是弹射座椅的发明者之一。

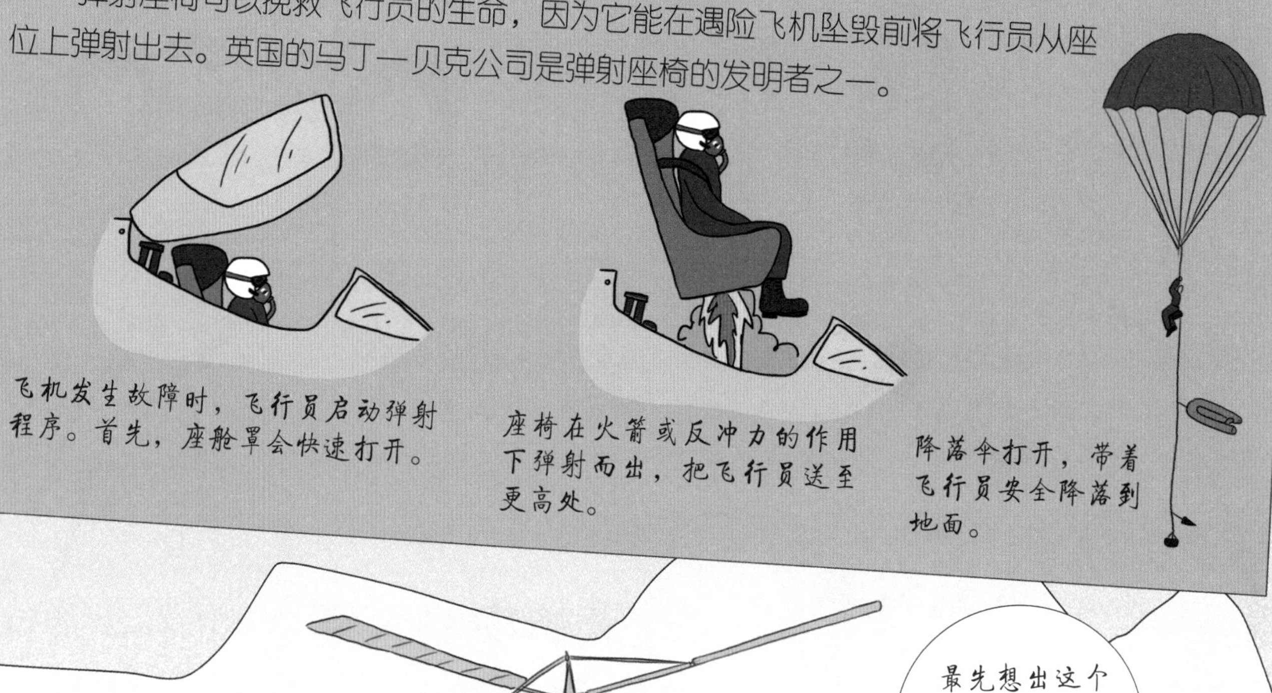

飞机发生故障时，飞行员启动弹射程序。首先，座舱罩会快速打开。
座椅在火箭或反冲力的作用下弹射而出，把飞行员送至更高处。
降落伞打开，带着飞行员安全降落到地面。

VS-300是第一架成功使用尾部螺旋桨的直升机。
最先想出这个点子的人是莱奥纳多·达·芬奇！

直升机

1907年，自行车制造商保罗·科尔尼进行了第一次载人直升机飞行，但他的试验机只飞离了地面几英尺。世界上第一架真正成功飞行的直升机是VS-300，是俄裔美国人伊戈尔·西科尔斯基在1939年设计的。

太空旅行

如果你想探索太空，喷气式飞机和内燃机可没法带你去，你得坐火箭。大型火箭自带燃料和氧气，动力足够强劲，能摆脱地球引力。太空旅行虽然只有几十年的历史，但科学家和科幻小说作家很久以前就提出了关于太空旅行的设想。

中国的“火箭”

火箭的基本原理是这样的：发动机向后排气产生推力，从而推动火箭向前。中国人发明火药时就掌握了这一原理。

人类登上月球

1971年，美国宇航员戴维·斯科特和吉姆·欧文乘坐阿波罗15号飞船执行任务，在月球上度过了三天时间。他们驾驶电动载人月球车（LRV）在月球表面行驶。

早在1903年，齐奥尔科夫斯基曾预言，人们会使用装有一次性助推器的多级火箭。

火箭科学

俄罗斯人康斯坦丁·齐奥尔科夫斯基（1857—1935）破解了火箭科学中的许多数学难题。虽然他自己从未造过火箭，但他的工作启发了后来的德国和美国航天先驱。

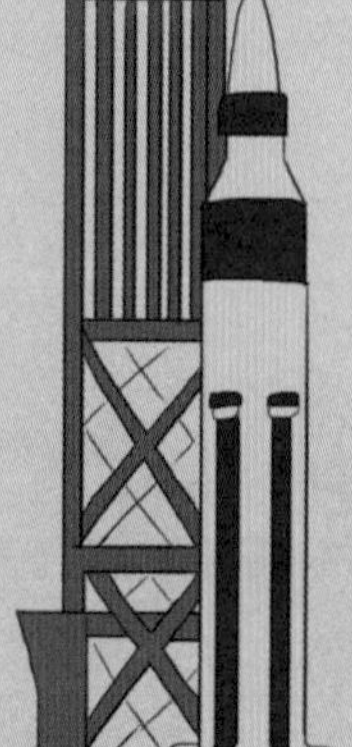

液体燃料火箭

1926年3月16日，美国科学家罗伯特·戈达德发射了世界上第一枚液体燃料火箭。火箭升空了12米，然后在罗伯特的姑姑埃菲的圆白菜地里坠毁。尽管如此，罗伯特·戈达德依然创造了历史。

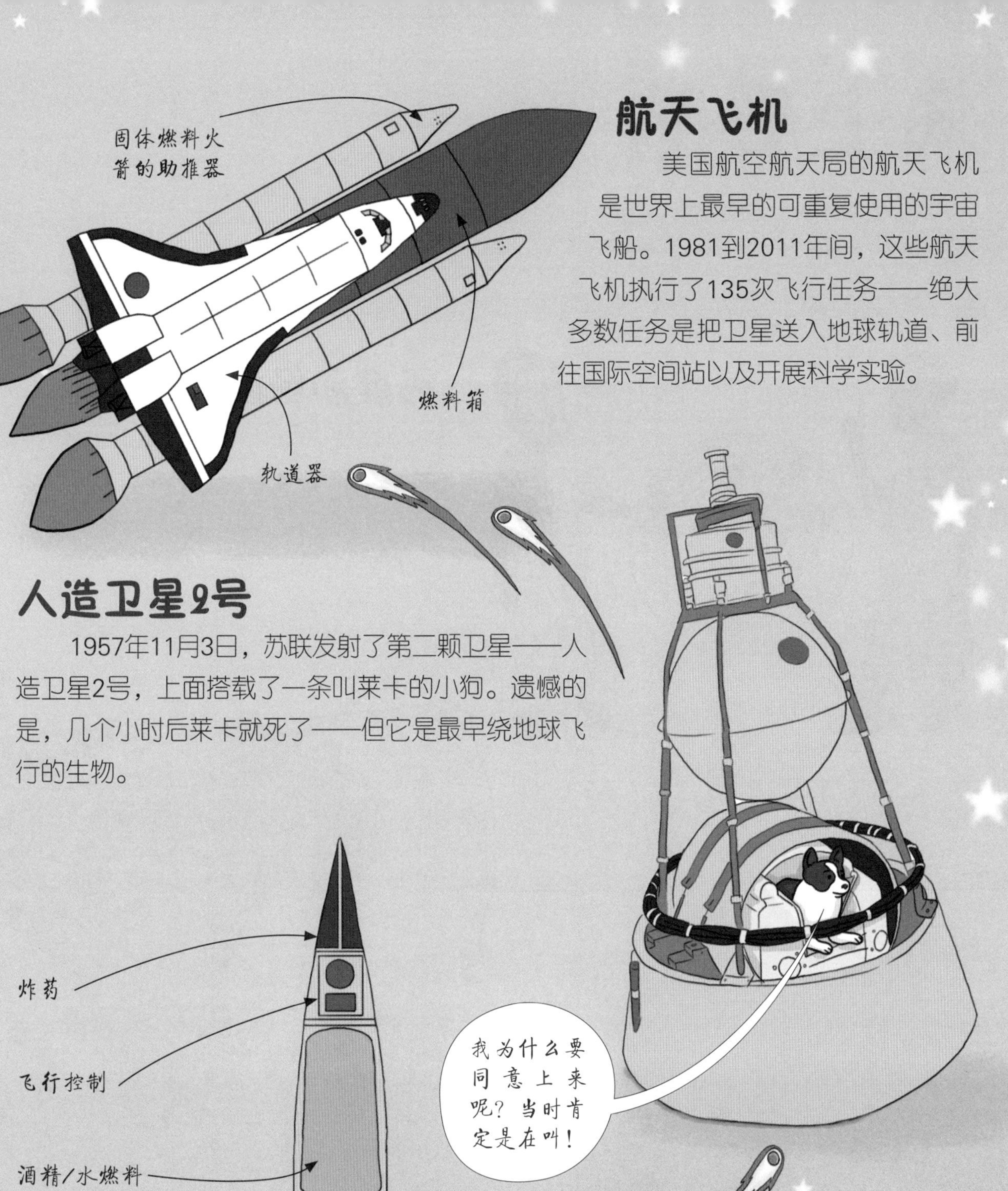

航天飞机

美国航空航天局的航天飞机是世界上最早的可重复使用的宇宙飞船。1981到2011年间，这些航天飞机执行了135次飞行任务——绝大多数任务是把卫星送入地球轨道、前往国际空间站以及开展科学实验。

人造卫星2号

1957年11月3日，苏联发射了第二颗卫星——人造卫星2号，上面搭载了一条叫莱卡的小狗。遗憾的是，几个小时后莱卡就死了——但它是最早绕地球飞行的生物。

V2火箭

第二次世界大战期间，维尔纳·冯·布劳恩为纳粹德国发明了“复仇武器”——V2火箭。后来，他利用自己的技术帮助美国发展了太空计划。

海上导航

出去旅行需要知道自己所在的位置和要去的地方，寻路的过程即为导航。在早期，陆地导航就够复杂了，海上导航更是难上加难——至少在地图、航海图、船只测速仪、指南针和其他导航辅助设备发明前是这样。

埃拉托色尼
喜帕恰斯
埃拉托色尼利用太阳投下的影子来计算地球的周长。真聪明！

纬度和经度

地图上有纬线（沿东西方向环绕地球）和经线（沿南北方向穿过南北两极）。古希腊人埃拉托色尼和喜帕恰斯“发明”了纬线和经线。

航海图

水手使用的海洋地图叫做航海图。13世纪，意大利水手绘制出了世界上第一幅“波托兰海图”，上面有罗盘线，为商人和探险家们标明了路线。

船只测速仪

船只测速仪通过水测量速度。最早的测速仪实际上是一段原木，上面系着一根绳，绳上打着一个个结。水手们把这段原木扔进水中，然后数有多少绳节从手里放了出去。所以船的速度单位为节。

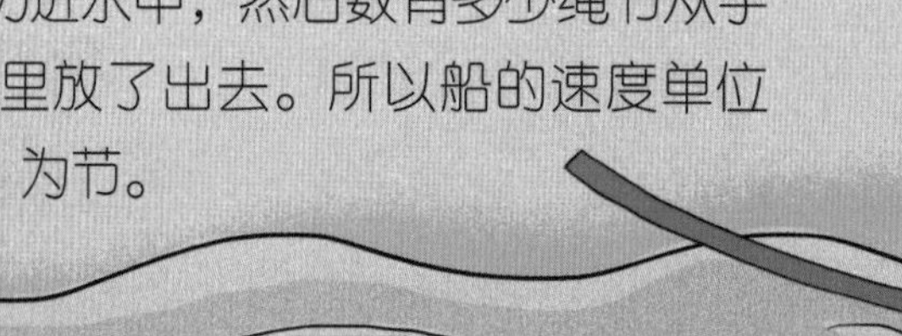

全球定位系统

现在水手们有了全球定位系统（GPS），该系统使用卫星数据生成非常准确的电子航海图。这一技术由美国国防部发明，并从1995年开始使用。

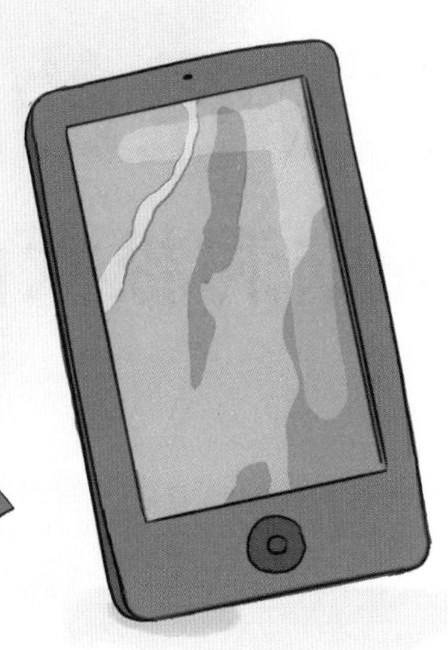

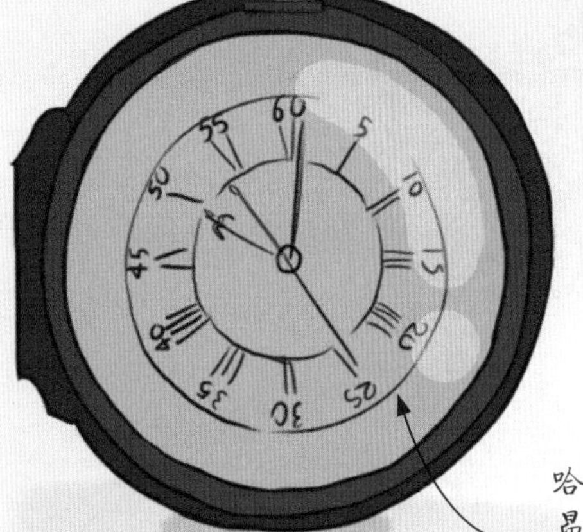

哈里森的航海钟价格高昂，大约相当于一艘船总造价的三分之一！

航海钟

如果你在海上能计时，就可以算出自己所处的位置。左图是1761年的H4航海钟，是英格兰钟表匠约翰·哈里森花了六年时间制作出来的。詹姆斯·库克船长带着一块H4航海钟开启了远到太平洋的著名航行。

！？

四分仪

水手们很早就会利用星星来导航。右图这位18世纪的水手正在使用戴维斯四分仪，通过目视太阳来测量太阳的角度。

最早的指南针

古代中国人发明了指南针——在铜盘上放一把磁勺。到十世纪时，中国人找到了磁化指针的办法，造出了更好的指南针来帮助中国商船在海上航行。

稀奇古怪的发明

（一些脑洞极大的交通工具）

我为车狂

世界上第一辆“汽车”是法国工程师尼古拉－约瑟夫·屈尼奥在1769年设计的，是一辆以蒸汽为动力的大型三轮汽车，比最早的实用型汽车早了100多年。他还制造了世界上第一起车祸——1771年，他发明的这辆车失控撞上了一堵墙。考虑到这辆车没有刹车，出车祸并不意外……

最不可思议的造车创意是什么？嗯……可能是1958年的福特核能汽车，这种车配有核反应堆。还好这种车没被造出来。

穿上扑翼

人们最早研究载人飞行时，许多先驱都试图模仿鸟类飞行。他们制造了扑翼机——一种有扑翼的飞行装置。有的扑翼机居然飞起来了，真是难以置信……

1869年，W. E.昆比先生发明的翼服获得了专利。幸亏他从来没有试过穿上它飞行。

1928年，乔治·怀特建造了上图这架以脚为动力的扑翼机，并驾驶它在美国佛罗里达州的一片海滩上飞行。这辆扑翼机的双翼每分钟扇动约80下。还不错！

“四不像”的发明

古希腊人曾告诫我们要“认识自我”。但有些人显然没听进去。看看那些四不像的飞机你就明白了。连飞机自己都不知道是要下潜，还是要在水中漂浮，是要在地上跑，还是要飞行。

1936年，苏联工程学专业的一名学生鲍里斯·乌沙科夫设计了下图这艘飞行潜艇，预计空中飞行时速为185千米，水中航行时速为5.5千米。1947年，鲍里斯的老板最终同意对它进行一次试运行，不过没有成功。

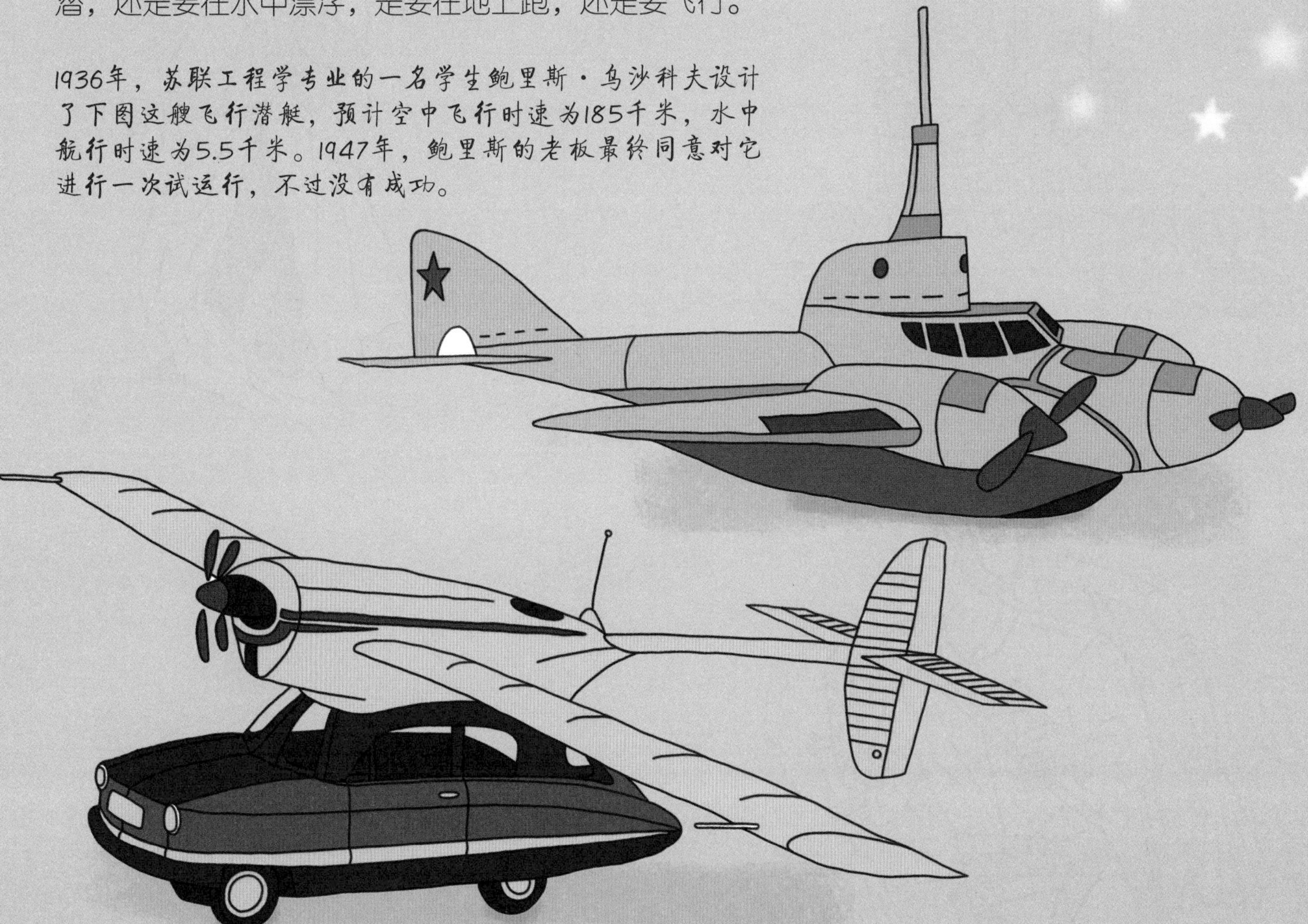

如上图所示，康维尔118型是一辆飞行汽车。1947年11月，试飞员驾驶它飞向高空时忘了检查燃料，随后飞机燃料耗尽，试飞员只好紧急迫降。1948年1月，飞行汽车再次起飞，但那时康维尔已经失去了兴趣，这个计划也就此搁浅。

勇于冒险的飞行家

今天，冒险家们利用悬挂式滑翔机、微型飞行器或翼服独自飞行，让飞行看上去很简单，甚至很安全……但是在过去，人类走了许多弯路才发明出了完美的个人飞行装备。

中国的火箭式烟火

传说中国有个名叫万户的航天先驱。他十分勇敢，在椅子上绑了一些火箭式烟火，然后将其点燃。有记载说万户飞向了高空，但也有记载说当时发生了大爆炸："当烟雾散去，万户和椅子都不见了，人们再也没有见过他。"

翼服

1912年，裁缝兼发明家弗朗茨·赖歇尔特发明了翼服，他希望这件翼服有和降落伞一样的功能。为了测试它的功能，赖歇尔特从巴黎埃菲尔铁塔一跃而下……结果坠地而亡。

喷气背包

你想有一个喷气背包吗？20世纪50年代，美国的贝尔公司发明了左图这个单人火箭腾越器——但它只飞行了20秒。

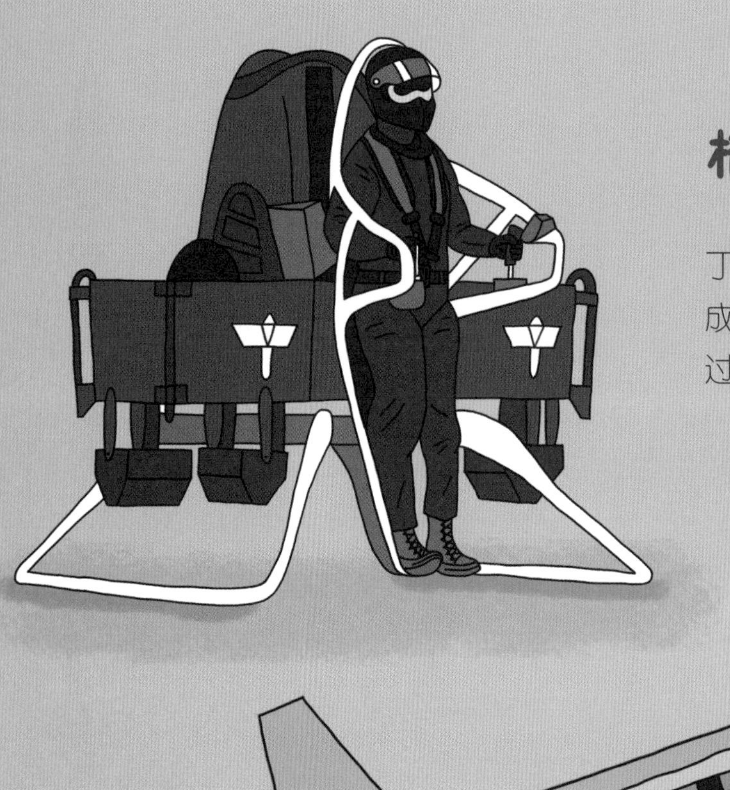

格伦·马丁

从1981年起，新西兰人格伦·马丁就一直在研究喷气背包。最后终于成功了……不过他的喷气背包造价超过了十万美元。

喷气式飞行翼

瑞士飞行员伊夫·罗西也设计了喷气式“飞行翼”，飞行最高时速可达300千米。2008年，他只花了13分钟就飞过了英吉利海峡。

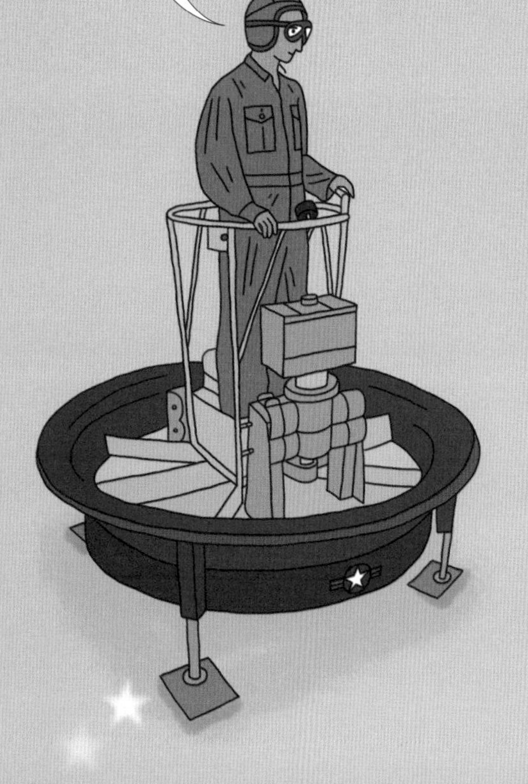

涵道风扇垂直起降试验机

涵道风扇垂直起降试验机是20世纪50年代中期专为美国军队设计的，设计者是飞机工程师查尔斯·齐默尔曼。但是这架试验机又小又慢，动力也很弱，最后只升空了十米左右。

索引

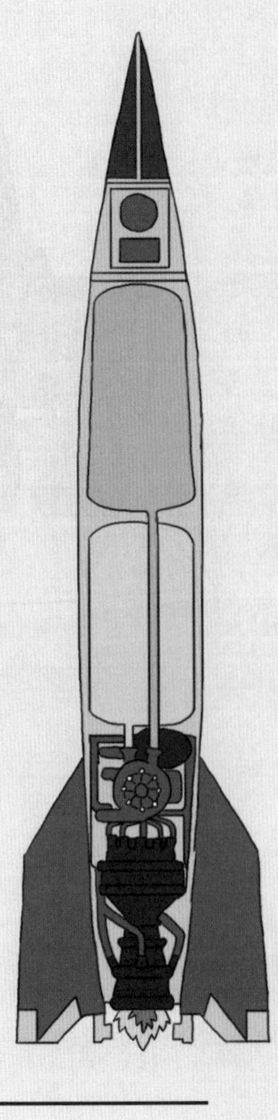

作者简介

马特·特纳出生于英国，20世纪80年代毕业于拉夫伯勒大学艺术学院，毕业后一直担任图片研究员、编辑和作者。他的书题材广泛，涉及博物学、地球科学和铁路等，并为百科全书和分册出版的丛书写过数百篇文章，从大象到抽象艺术无所不包。他现在和家人住在新西兰奥克兰附近，他还是当地海岸警卫队的志愿者，平时也涉猎工艺品的制作。

绘者简介

萨拉·康纳生活在英格兰美丽的乡村，栖身于一座可爱的小木屋里，有几只狗和一只猫做伴。她平时用画笔描绘身边的世界。她总能从大自然中获得灵感，这对她的很多作品都有影响。萨拉之前用钢笔和颜料画插图，但近些年，她开始在电脑上绘画，因为这更适应今天的产业发展。然而，她还是喜欢时不时拿出水彩颜料来，画一画自己花园里的鲜花！

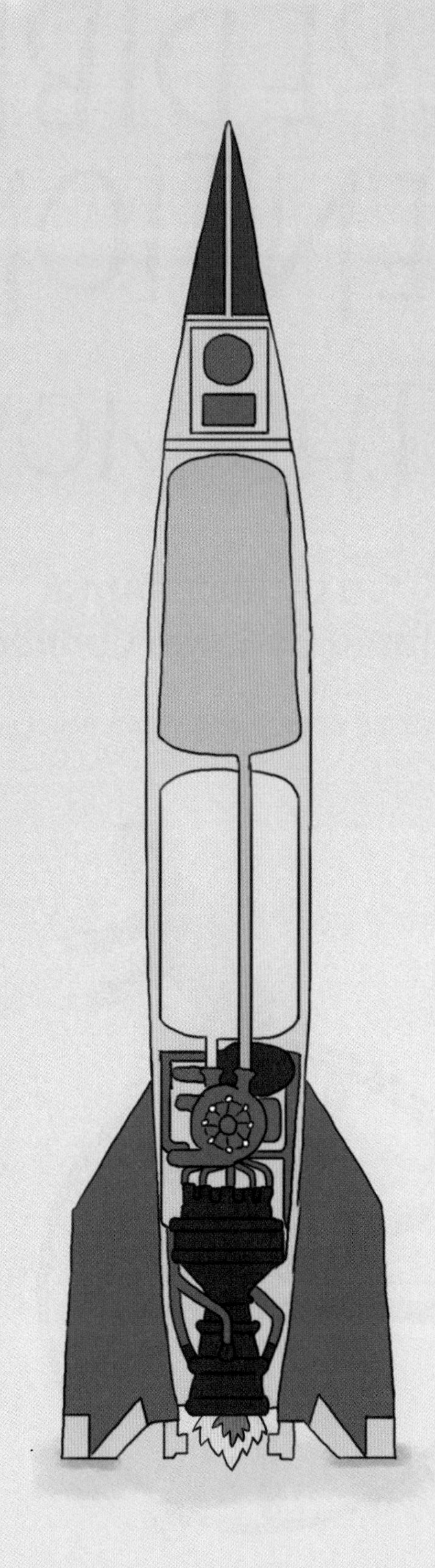

Chart the progress of roadworthy vehicles from chariots to the Volkswagen Beetle, and seaworthy craft from dugout canoes to aircraft carriers. Follow the railway revolution and the conquest of the air from simple Chinese kites to jet aircraft. And then outer space beckons... You'll find that

- Karl Benz's first 'car' had three wheels
- a triangular sail transformed sea voyaging
- steam locos chugged into life over 200 years ago – at 3.9 km/h
- the idea for modern space rockets dates from 1903

and so much more. And don't miss the wild designs that didn't get off the ground (like the nuclear-powered car and flying submarine), and those that did, such as daredevil jetpacks and wingsuits...

In 2015, a Japanese maglev (magnetic levitation) train reached a record 605 km/h in a test.

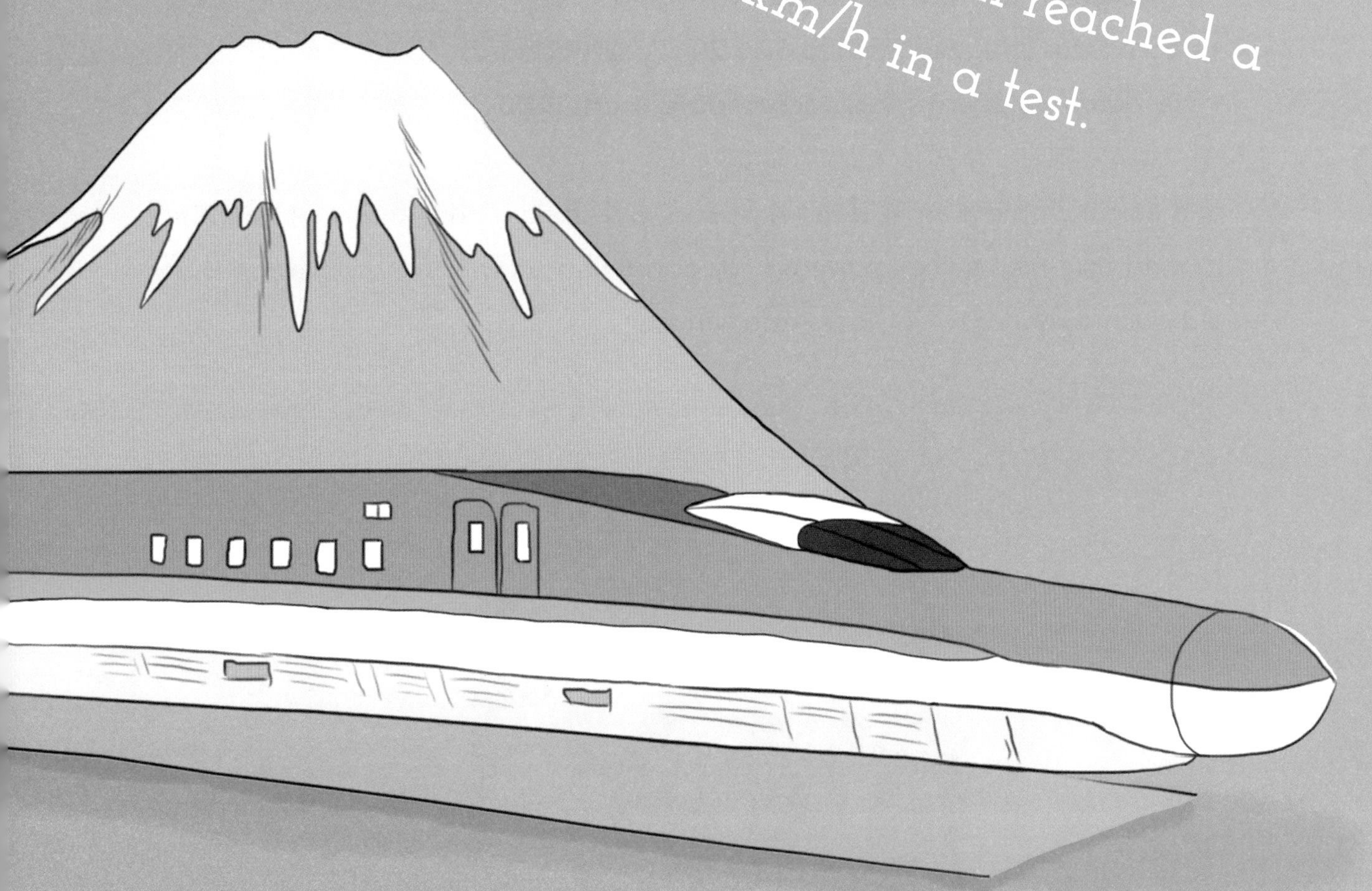

CONTENTS

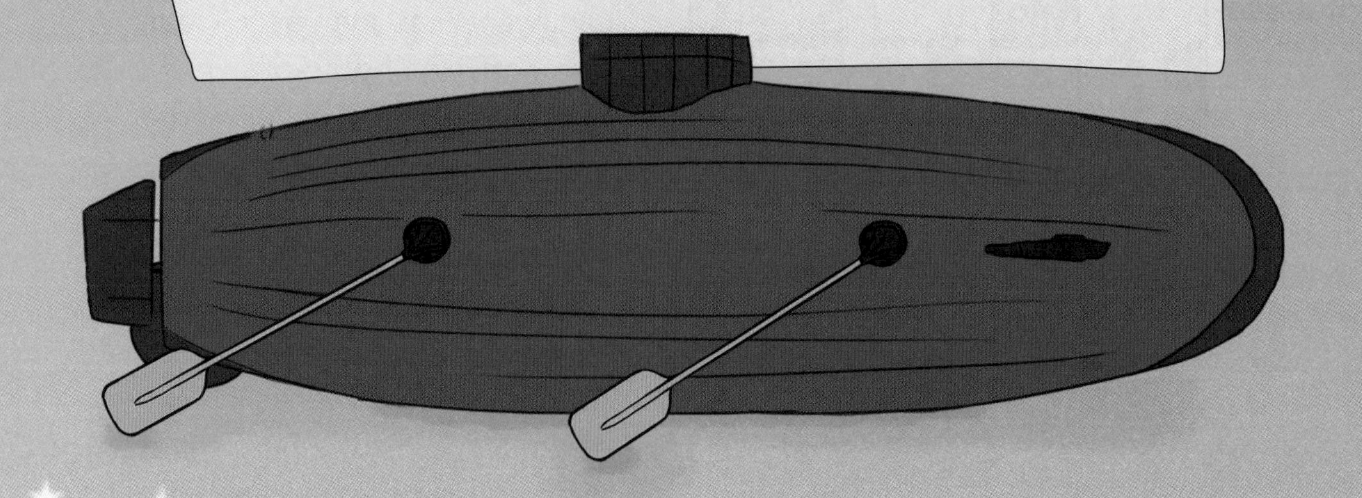

OFF WE GO!

We humans are natural wanderers, but for tens of thousands of years our ancestors moved on foot. They had no other means of transport – not, at least, until they had tamed wild animals. Early humans domesticated the horse, for instance, about 6,000 years ago. Many years later, they invented the wheel – which led to carts and chariots. They also tamed the wind to power sailing vessels so they could cross oceans.

Spoked wheel, 2,000 BC

The most dramatic revolutions in transport have come in the last two centuries, with the development of the steam engine and internal combustion engine. They powered boats, trains, cars and finally aircraft. They brought benefits – speed, comfort – but also problems, such as pollution. Did you know, by the way, that cars were first seen as a 'clean' alternative to horse-drawn transport? That's because the tons of horse poo dropped in city streets attracted flies, which spread disease. Now, of course, we're choking on exhaust fumes, and looking for alternative powers such as electricity. (Even the Bate car, invented by British chicken farmer

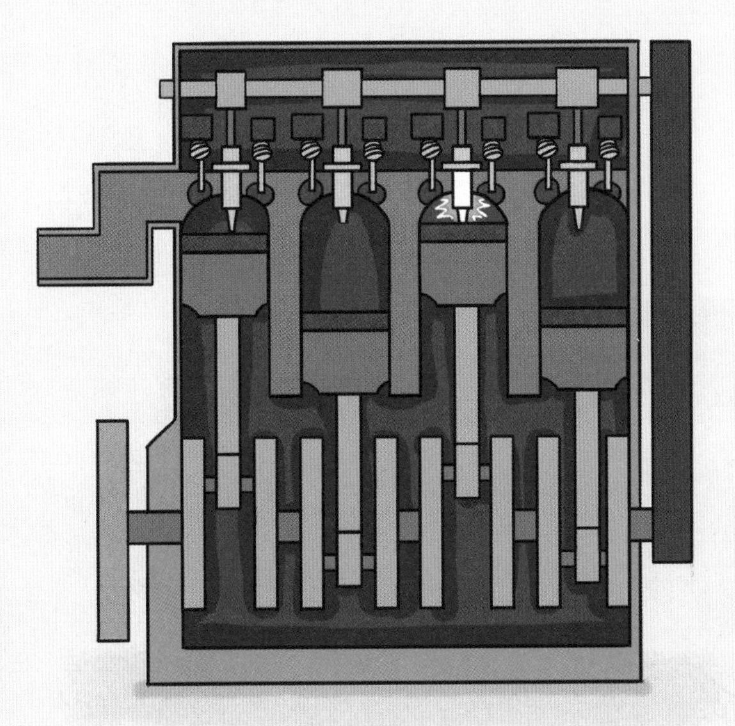

Internal combustion engine, from 1850s

Harold Bate in 1971, looks like a healthier idea: it ran on animal poo!)

The first aviation pioneers took inspiration from birds, but their flapping flying machines were a disaster. It took the Wright brothers' historic flights of 1903 to unlock the mysteries

Dugout canoe, 6,000 BC (or earlier)

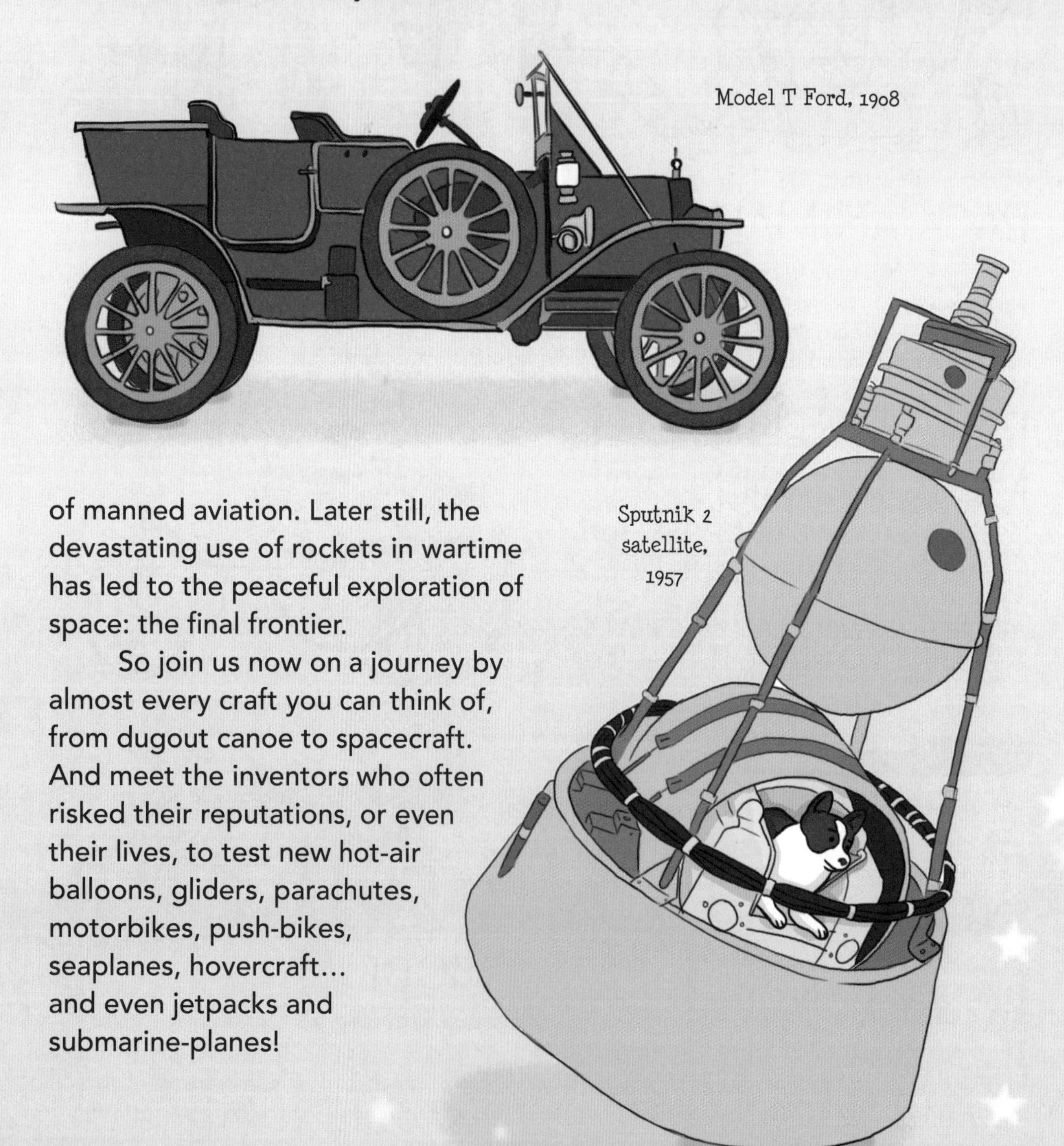

Model T Ford, 1908

Sputnik 2 satellite, 1957

of manned aviation. Later still, the devastating use of rockets in wartime has led to the peaceful exploration of space: the final frontier.

So join us now on a journey by almost every craft you can think of, from dugout canoe to spacecraft. And meet the inventors who often risked their reputations, or even their lives, to test new hot-air balloons, gliders, parachutes, motorbikes, push-bikes, seaplanes, hovercraft... and even jetpacks and submarine-planes!

ON THE ROAD

The invention of the wheel revolutionized transport and technology, but it came fairly late in human history – maybe around 3,000 BC, long after humans had invented spears, flutes and pottery. In fact the idea came from the potter's wheel. The reason it took them so long is probably because there are no wheels in nature. This was a completely human invention.

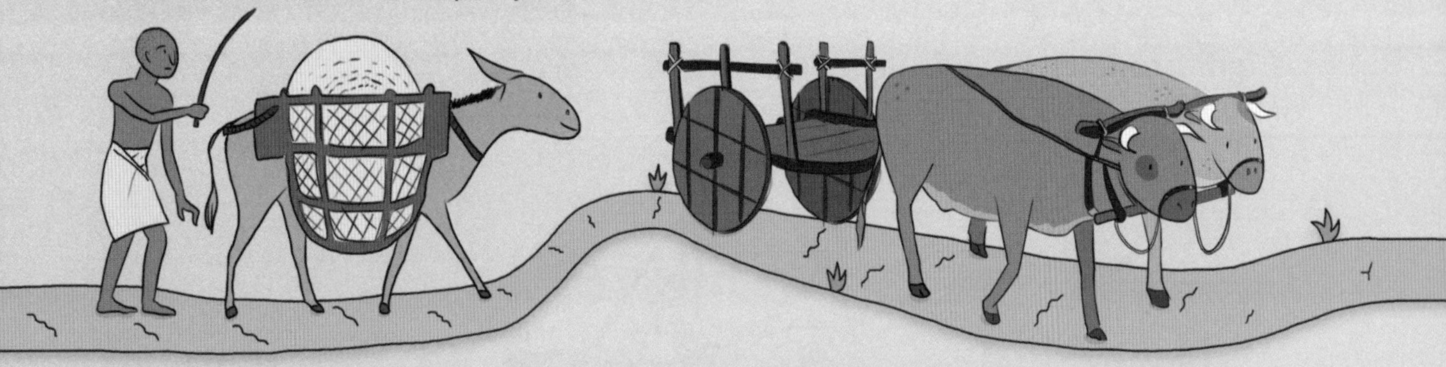

PACK ANIMAL

Before we had wheels, we used pack animals. The onager or wild ass served as a beast of burden in ancient Sumer (modern Iraq).

FIRST WHEEL

The earliest wheels were solid planks, shaped and joined together. Carts like this one appeared in northern Europe in about 2,500 BC. The cart is drawn by domestic cattle.

BENZ PATENT-MOTORWAGEN

In 1885, German inventor Karl Benz (later of Mercedes-Benz fame) produced the three-wheeled Patent-Motorwagen. His wife, Bertha, took it to visit her old mum, and on the way she invented brake lining.

ROLLS-ROYCE

The Rolls-Royce 40/50 Silver Ghost appeared in 1907. A car magazine called it 'the best car in the world'; it was also the most expensive.

MODEL T

American Henry Ford designed his Model T car in 1908. In 1913, he invented the production line: cars were built in a strict sequence – three cars per minute!

SPOKED WHEELS

The Andronovo people, riding their chariots on the Eurasian plains 4,000 years ago, are thought to have invented spokes. Spokes made wheels lighter, bigger and faster.

STAGECOACH

The coach was invented in Koc (pronounced as 'kotch'), Hungary. In 17th-century Europe, long journeys were made by stagecoach. Early models gave a bumpy ride, but by 1800, coaches were fast and comfy, with sprung suspension and brakes.

STEAM ENGINE

Steam engines, used in boats from the 1780s, and later in railway locos, also began to power road vehicles. The 'Obedient' was a steam carriage made by Frenchman Amédée Bollée in 1875.

HIPPOMOBILE

In the late 1850s, Étienne Lenoir of Belgium invented a gas-powered engine, and later used it in his Hippomobile (meaning 'horse car', not 'hippo car'). Its top speed was 3 km/h – slower than walking pace.

Need a push?

Volkswagen means 'people's car'. By 2003, more than 21 million Beetles had been sold.

BEETLE

Porsche of Germany makes luxury sportscars. But back in the 1930s, Ferdinand Porsche also designed the very affordable Volkswagen Beetle – named after its bug-like shape.

Hmm, what shall I call it?

THE IC ENGINE

The internal combustion (IC) engine is fuelled by petrol, diesel or gas. Fuel explosions in the cylinders give out energy, which creates movement to turn the wheels. The IC engine appeared over 150 years ago, with many improvements added since.

Camshaft

Spark plug

Valves

Piston

Crankshaft

Cylinder

THE FOUR-STROKE CYCLE

1. Fuel/air mix enters cylinder.

2. Piston rises, squeezing fuel/air mix.

3. Spark plug ignites fuel, forcing cylinder down.

4. Piston returns, pushing exhaust (burnt gas) out.

CARBURETTOR

Fuel is vaporized inside the carburettor (right, which works a bit like an old-fashioned perfume spray). The fuel mist then enters the cylinder heads, where it is ignited by the spark plugs. The first effective carburettors were invented in the 1880s.

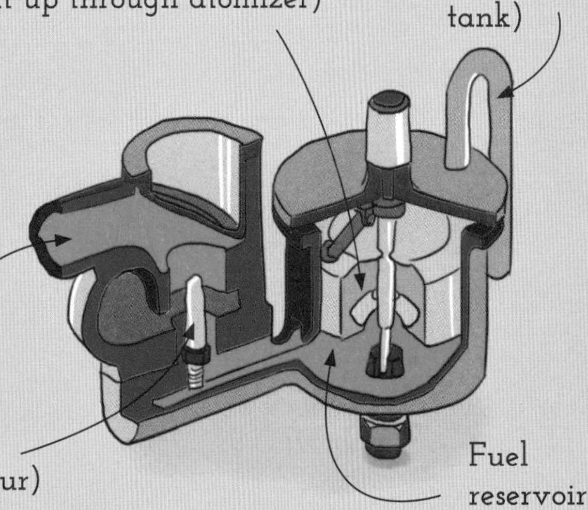

GAS ENGINE

Remember Étienne Lenoir and the hippo car? This is his 1850s engine. The fuel was coal gas, which in those days was used for street lighting. But around that time, we began converting crude oil into paraffin, kerosene and petrol. Petrol became the preferred engine fuel.

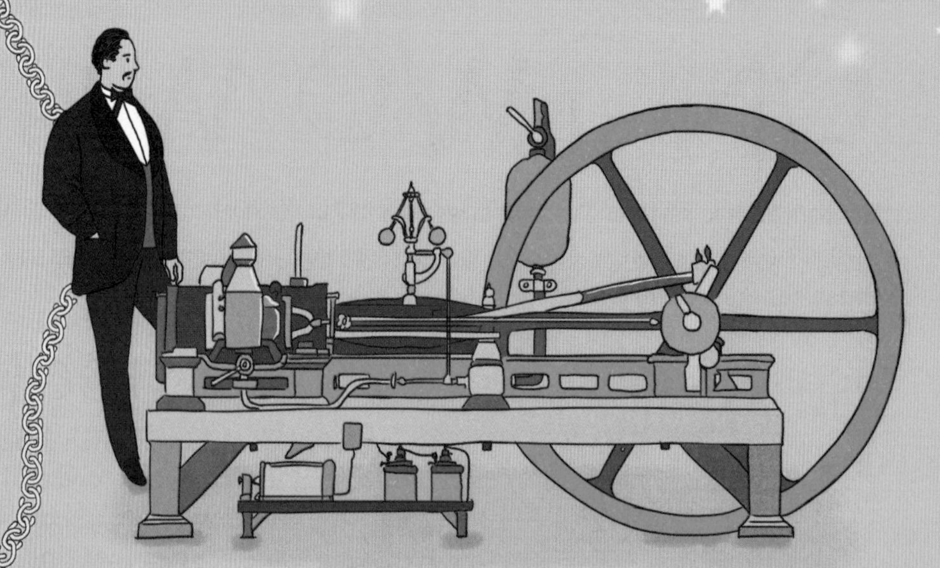

I shall call it... Diesel!

DIESEL ENGINE

In 1892–97, German Rudolf Diesel invented and improved the engine that bears his name. His first diesel engine was seven times more efficient than steam engines: it produced more power from less fuel.

A V8 engine has eight cylinders. They are arranged in a V-shape, with two blocks of four.

V8 ENGINE

Frenchman Léon Levavasseur invented the V8 engine in 1902, naming it the 'Antoinette' (after the daughter of his sponsor). It was mostly used in planes. This 1905 Darracq racing car had a 22.5-litre V8. It broke the land speed record!

ON TWO WHEELS

It's hard to imagine a world without bikes, isn't it? But they were only invented about 200 years ago. The first 'bone-shakers' were wood and iron, and very hard on the bum. Motorbikes, too, began as very crude machines, around 150 years ago.

DRAISINE

The wooden Draisine of 1817 was named after its German inventor, Karl von Drais. It had no pedals, so he called it a 'running machine'.

VELOCIPEDE

The French Michauline, or velocipede, of the 1860s was designed by Pierre Lallement and Pierre Michaux. It had no tyres but the seat was sprung.

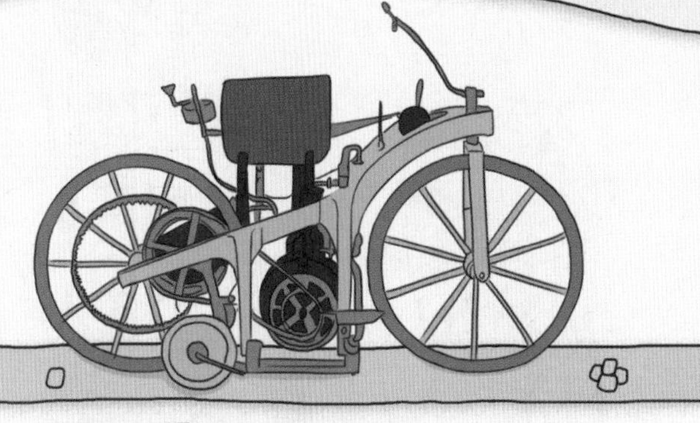

PETROL CYCLE

Englishman Edward Butler invented the three-wheeled Petrol Cycle in 1884. But its top speed of 16 km/h excluded it from built-up areas; the city speed limit was 3 km/h!

REITWAGEN

The Reitwagen, made by Gottlieb Daimler and Wilhelm Maybach, was the first true motorcycle, in 1885. But its seat caught fire. Ouch!

EARLY MOTORBIKE

Hildebrand & Wolfmüller, Germany, 1894 – one of the first reliable motorbikes

TRIUMPH

The Triumph Type H, England, 1915, used by messengers in World War I

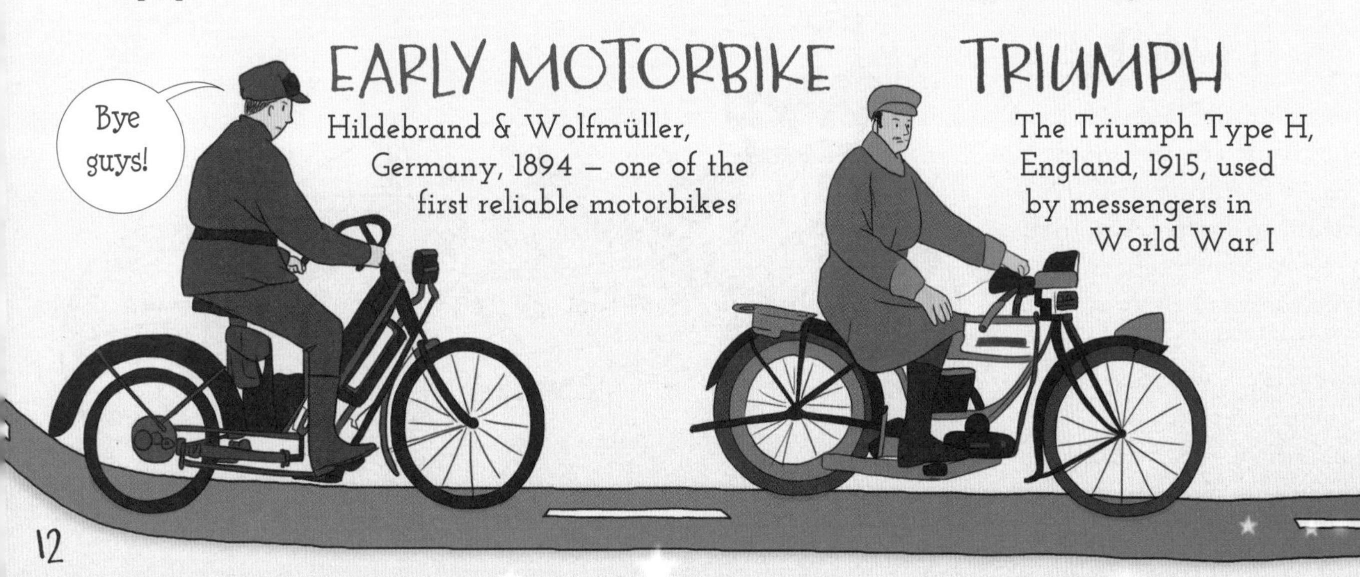

PENNY-FARTHING

The 'ordinary' of the 1870s was also called a 'penny-farthing', because its wheels looked like two coins – one small, one large.

SAFETY BIKE

Penny-farthings were very hard to ride, and crashed a lot. They were replaced later by 'safety bicycles'. This is the Rover safety bicycle of 1885, made by Englishman John Starley.

HELMET

The lid that saves lives... After famous British soldier T. E. Lawrence died in a motorbike crash in 1935, doctors began calling for the use of crash helmets. In America, racing driver Rory Richter was inspired to launch the Bell helmet brand in the 1950s, after a friend died in a crash.

MICHAUX

The Michaux steam-powered bike of 1867 was basically a Michauline push-bike with a steam engine attached.

VESPA

Vespa MP-6 'moped', Italy, 1946 – a motor scooter for everyone

HONDA

Honda CB750, 1969 – one of the first modern, fast Japanese bikes

THE FIRST BOATS

Since paddling the very first fallen log across a creek, we humans have been steadily improving watercraft design. Sails were first used by the Mesopotamians some 5,000 years ago, but the invention of the triangular lateen sail, about 1,800 years ago, enabled mariners to put to sea in almost any wind. As shipping evolved, so did sea power and trade.

Row, row, row your boat...
You guys are oarsome!
Beak

DUGOUT CANOE

From at least 8,000 years ago, dugout canoes served worldwide as simple watercraft. First, pick a long, straight tree trunk, then chop (or burn) the wood from the centre. Now take your oar...

EGYPTIAN BARGE

This Egyptian barge, built 4,500 years ago, possibly carried the body of King Khufu to his tomb at Giza. There, it was sealed into a pyramid for the dead pharoah to use in the afterlife.

BIREME

Around 800-700 BC, the Phoenicians invented the bireme, a ship with two banks of oars on each side. Some biremes were used for trade, others for battle. War galleys had a sharp 'beak' for ramming enemy ships.

AIRCRAFT CARRIERS

Modern aircraft carriers are gigantic to allow aircraft to take off. Some have measured up to 333 m long and carried up to 90 aircraft. This ship is painted in dazzle camouflage.

1826 propeller designed by Czech inventor Josef Ressel

PROPELLERS

Ships' propellers were invented in the early 1800s. Paddle wheels are much older; the Romans even had paddle ships powered by cows!

WIND POWER

The caravel was a 15th-century Portuguese ship whose triangular lateen sails allowed it to sail in almost any wind direction. This example is the *Matthew*, in which explorer John Cabot crossed the Atlantic in 1497.

GIANT JUNKS

From 1405 to 1433, Chinese traders and explorers mounted sea expeditions as far west as Africa. Under the command of Admiral Zheng He, they sailed in huge fleets of sailing ships known as junks, some 122 m long – among the largest wooden ships ever built. They were often called 'treasure ships'.

WATER WONDERS

Why travel on the water when you can go under it? The history of the submarine dates back 400 years – but you wouldn't want to try some of those ancient leaky tubs! Or perhaps a hovercraft is what really floats your boat?

OUTBOARD MOTOR

This 1902 French outboard motor made steering tricky – like using a giant egg-beater! The first serious outboard was made in 1907 by Norwegian-born Ole Evinrude.

HOVERCRAFT

In 1959, Christopher Cockerell crossed the English Channel in his first full-size hovercraft, the SR.N1. His design borrowed ideas from hovercraft research dating back to the 1870s.

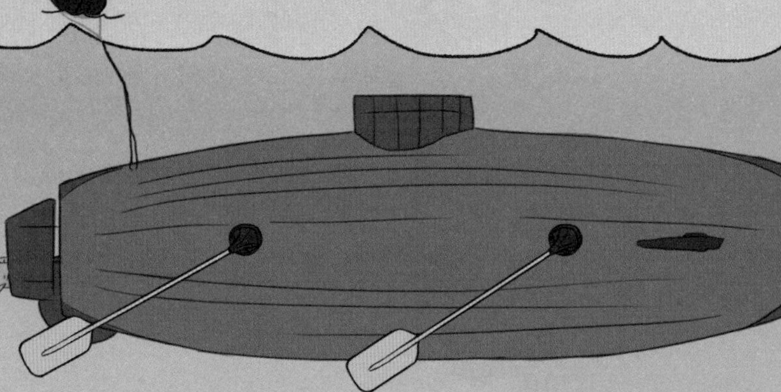

SUB WITH OARS

Cornelius Drebbel was a brilliant Dutchman whose many inventions included a submarine with oars. In 1624, he showed his latest sub to King James I of England, and even took the monarch for a ride in the River Thames!

TURTLE

In 1775, American David Bushnell designed the *Turtle*, a one-man submarine. He wanted to use it to place explosives secretly on British ships during the Revolutionary War.

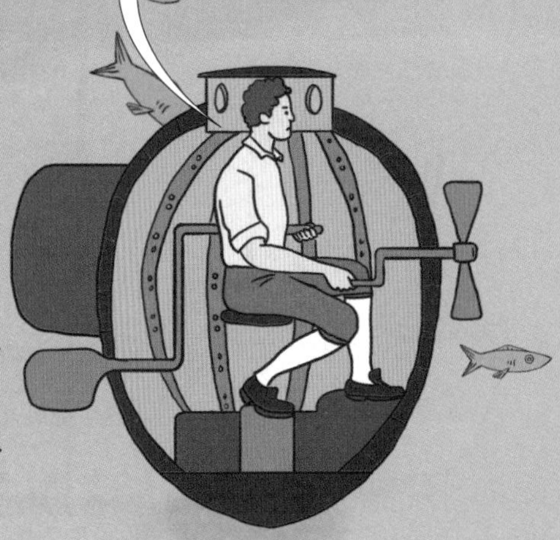

GROUND EFFECT

Ground-effect aeroplanes float several metres above water on a cushion of trapped air. This huge Lun aeroplane, built in Russia in the 1960s, measured over 90 m long. It was unstable, though, so the idea didn't really 'take off'.

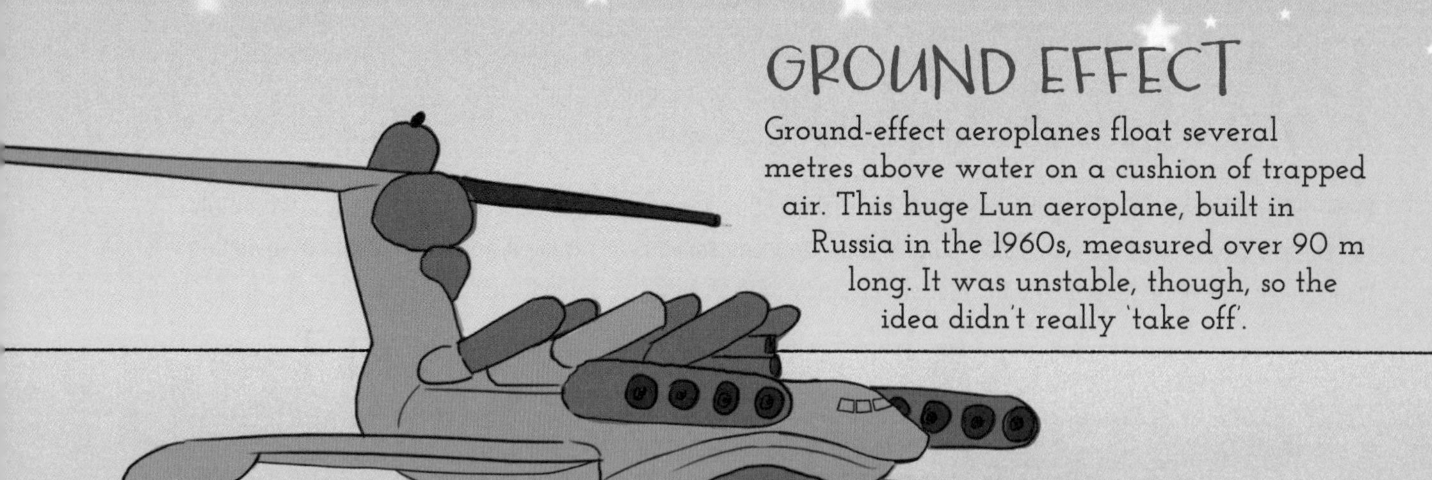

JETBOATS

Jetboats were invented by New Zealander Bill Hamilton in the 1950s. The engine scoops up water from below and pumps it out the back to push the boat forward.

HOLLAND 1

This is the British Royal Navy's *Holland 1* submarine of 1901, designed by Irish engineer John Holland. She sank off the British coast in 1913 while being towed, and was lost. In 1983 she was recovered and restored for museum display.

Holland 1 measured nearly 20 m long and was armed with one torpedo tube.

RAIL TRAIL

Powerful, high-pressure steam engines, invented in the 1800s, soon found their way onto the railways – although the earliest locomotives were very slow puffers! Since then, diesel and electric locos, and even 'magnetic levitation' trains, have drawn the railway into the modern era, providing us with fast, smooth, mass transport.

We are catching up with the snail!

STEAM LOCOS

Englishman Richard Trevithick developed the high-pressure steam engine. He trialled it in 1804, in the first loco-hauled railway trip: nearly 16 km at an average 3.9 km/h.

ROCKET

The 1829 Rainhill Trials were held to select a loco design for the new Liverpool and Manchester Railway line. The winner was *Rocket*, built by George and Robert Stephenson. *Rocket* was very modern for its day.

BULLET TRAIN

Today's Shinkansen 'bullet trains' of Japan are capable of going much faster than their 320 km/h speed limit. In China, maglev (magnetic levitation) trains float on a magnetic field at speeds of up to 430 km/h.

An American diesel-electric of the 1920s, made by General Electric

DIESEL-ELECTRIC

Diesel-electric (DE) locos use a diesel engine to power a generator, which drives the electric motor that moves the train. When DEs replaced smoky old steam trains in city areas, it helped keep laundry clean!

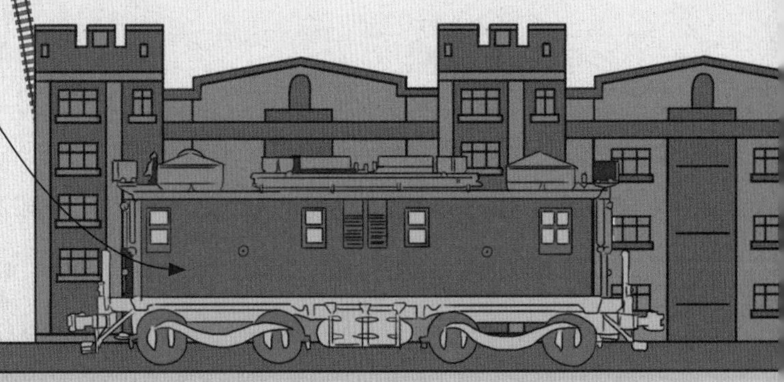

ELECTRIC TRAIN

In 1879, German inventor Werner von Siemens built this little electric train – the world's first. (The next year, he built the first electric elevator; and the year after that, the first electric tramway, in Berlin.)

Ermm, it's a little bit small.

Cor, blimey! It's not 'alf smoky in 'ere!

METROPOLITAN LINE

The world's first underground railway was the Metropolitan Line, which opened in London in 1863. Its steam locos made for a smoky ride (cough), so they were later replaced with electric trains.

Choo-choo coming through!

MALLARD

The British *Mallard*, designed by Sir Nigel Gresley in 1938, still holds the world steam speed record of just over 202 km/h. Its beautiful streamlining creates a sleek, slippery shape, keeping wind resistance low.

FIRST FLIGHT

You look at the birds and you want to fly too, right? You're not the first. The earliest attempts go back as far as ancient China. But it was the development of hot-air balloons just over two centuries ago, and much later the dawn of powered, controlled flight, that turned human dreams of flight into a reality.

CHINESE KITE

As much as 3,000 years ago, Chinese 'aviators' flew strapped to big kites. Later, sailors would send manned kites up into the air to test the prospects for a good voyage.

Hey, you down there... don't let go of that string!

LEONARDO'S MACHINE

The Italian artist-genius Leonardo da Vinci (1452-1519) invented a flying machine inspired by bats and birds, though he never actually built it. (He also invented helicopters – on paper, at least!)

FLOATPLANE

Floatplanes, or seaplanes, are aircraft that can take off from water, and land on it, too. They date from 1905, when French aircraft pioneer Gabriel Voisin piloted a float-glider over the River Seine. It was towed by a speedboat.

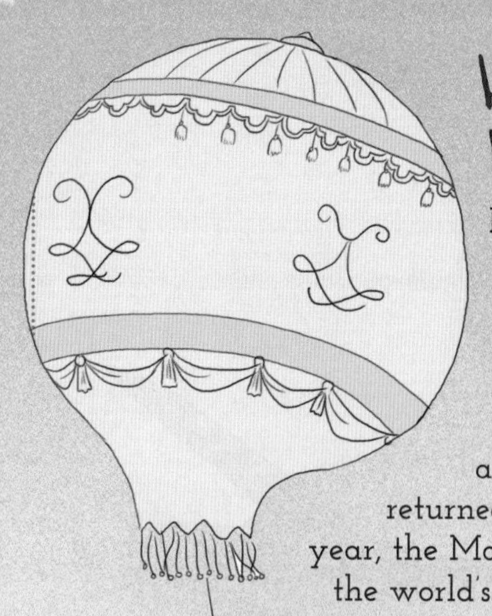

HOT-AIR BALLOON

France, 19 September 1783: the two Montgolfier brothers launch a hot-air balloon carrying a sheep, a duck and a rooster. The animals returned safely. Later that year, the Montgolfiers launched the world's first manned flights.

Baa... ooh-ah!

GAS BURNER

The Montgolfiers lit a fire under their balloons to lift them. In 1955, American Ed Yost introduced the on-board gas burner. In 1978, Yost's *Double Eagle II* made the first balloon crossing of the Atlantic Ocean.

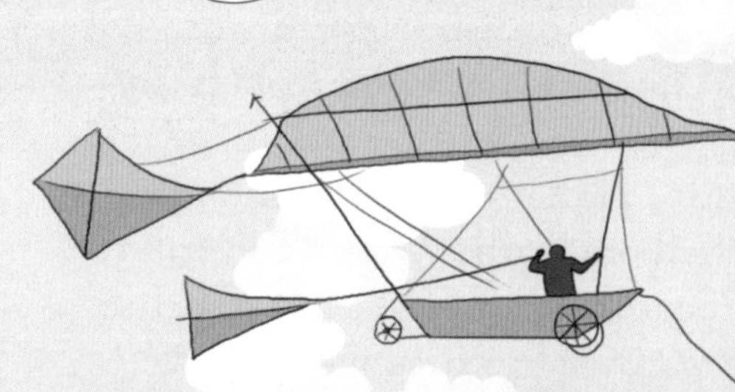

GLIDER

British inventor George Cayley sent a glider up in 1853, piloted by one of his servants (who, luckily, came down again unharmed).

AIRSHIP

This huge airship, which first flew in July 1900, was named after its German inventor, Count Ferdinand von Zeppelin. His first Zeppelin, the *LZ1*, measured 128 m – nearly three times the length of a modern jumbo jet.

AIR POWER

Next time you're in a plane, it'll probably be a comfy jet airliner with in-flight food, and maybe also movies. But only a century ago, planes were flimsy craft of 'stick, string and fabric', with little more than hope and courage to keep them in the air.

JET AIRCRAFT

The first jet planes also appeared in World War II. They were invented by Hans von Ohain in Germany and Frank Whittle in Britain. The Heinkel He178, powered by a von Ohain engine, made the world's first jet flight in 1939.

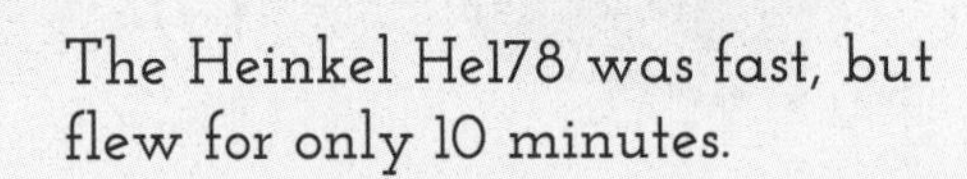

The Heinkel He178 was fast, but flew for only 10 minutes.

Spitfire. Lives up to its name. Better than Shrew, eh.

SPITFIRE

The British Supermarine Spitfire, designed in 1936 by Reg Mitchell, was a famous World War II fighter plane with wing-mounted guns. Mitchell originally wanted to call it 'Shrew' or 'Scarab', and didn't like 'Spitfire'. More than 20,000 Spitfires were built; about 50 still fly today.

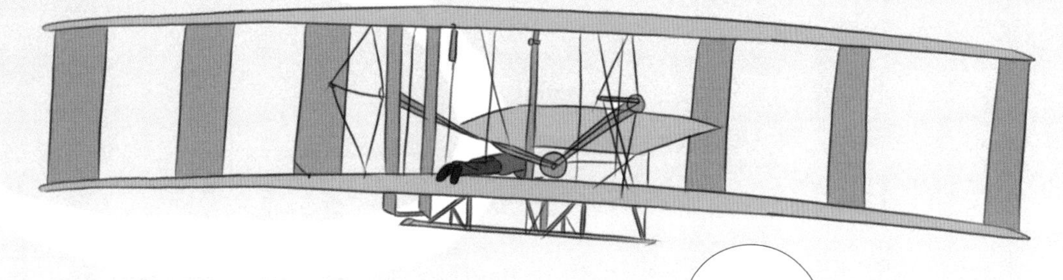

It flies!

WRIGHT BROTHERS

Wilbur and Orville Wright made the world's first powered flights at Kitty Hawk, North Carolina, in 1903. They had begun building gliders, but their first powered plane, the *Flyer 1,* carried a 12-horsepower motor.

CONCORDE

The beautiful British-French Concorde served as an airliner between 1976 and 2003. It carried up to 128 passengers at over twice the speed of sound. The long nose could be 'drooped' on landing so that the pilot could see over it!

EJECTION SEAT

The ejection seat saves pilots' lives by throwing them from a stricken plane before it crashes. Its inventors include the British company Martin-Baker.

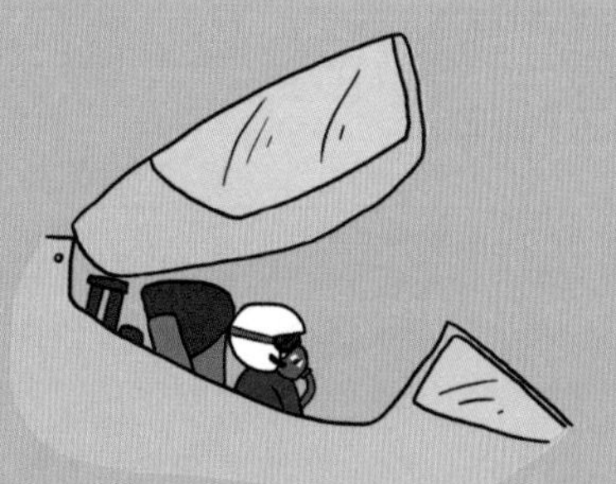

If a plane gets into trouble, the pilot triggers the ejection sequence. First, the canopy opens rapidly.

The seat is quickly blown out by rocket or jet power, lifting the pilot high above the plane.

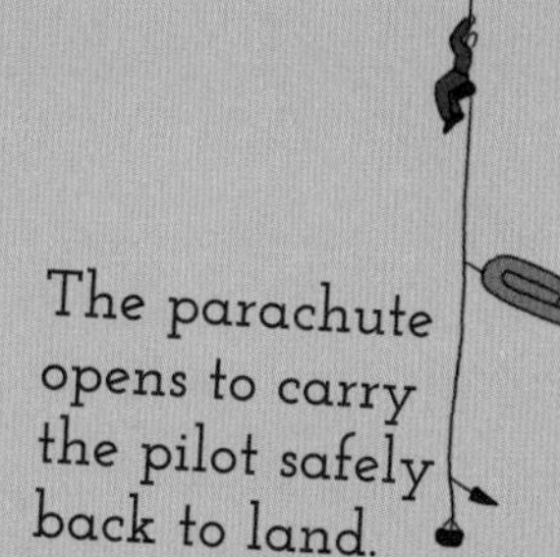

The parachute opens to carry the pilot safely back to land.

The VS-300 was the first helicopter to use a tail rotor successfully.

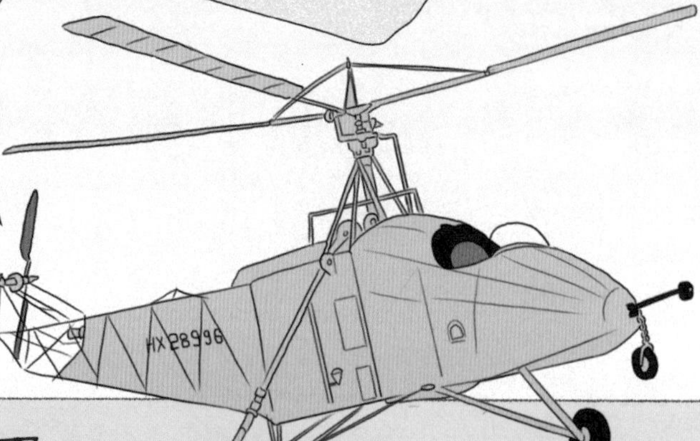

Leonardo da Vinci thought of that first!

HELICOPTER

In 1907 Paul Cornu, a bicycle maker, made the first manned helicopter flight, but his experimental craft rose only a few feet. The first truly successful heli was the VS-300, designed in 1939 by Russian-American Igor Sikorsky.

SPACE TRAVEL

If you want to explore space, jets and IC engines are not up to the job. You need a rocket. Large rockets are powerful enough to escape Earth's gravity, and carry all their own fuel and oxygen. Space travel is only a few decades old, but was predicted much earlier by scientists and sci-fi fantasy writers.

CHINESE ROCKET

Basic rocket science is this: the engine exhaust pushes backwards to push the rocket forwards. The Chinese grasped this when they invented gunpowder.

MEN ON THE MOON

On the Apollo 15 mission (1971), US astronauts David Scott and Jim Irwin spent three days on the Moon. They travelled the surface in this electric Lunar Roving Vehicle (LRV).

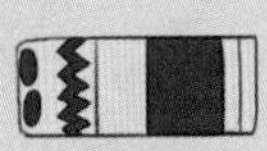

Back in 1903, Tsiolkovsky predicted the use of multi-stage rockets with disposable booster sections.

ROCKET SCIENCE

Russian Konstantin Tsiolkovsky (1857–1935) worked out a lot of the tricky maths in rocket science. Although he himself never built a rocket, his work was an inspiration for later pioneers in Germany and America.

LIQUID-FUEL ROCKET

On 16 March 1926, American scientist Robert Goddard launched the world's first liquid-fuel rocket. It rose to 12 m, then crash-landed in his Aunt Effie's cabbage patch. But he had made history.

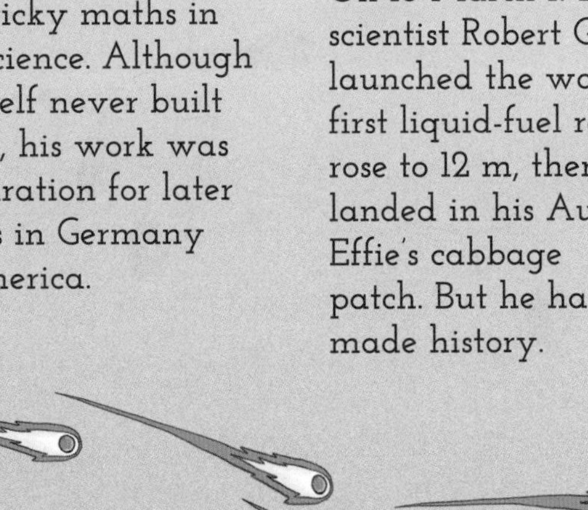

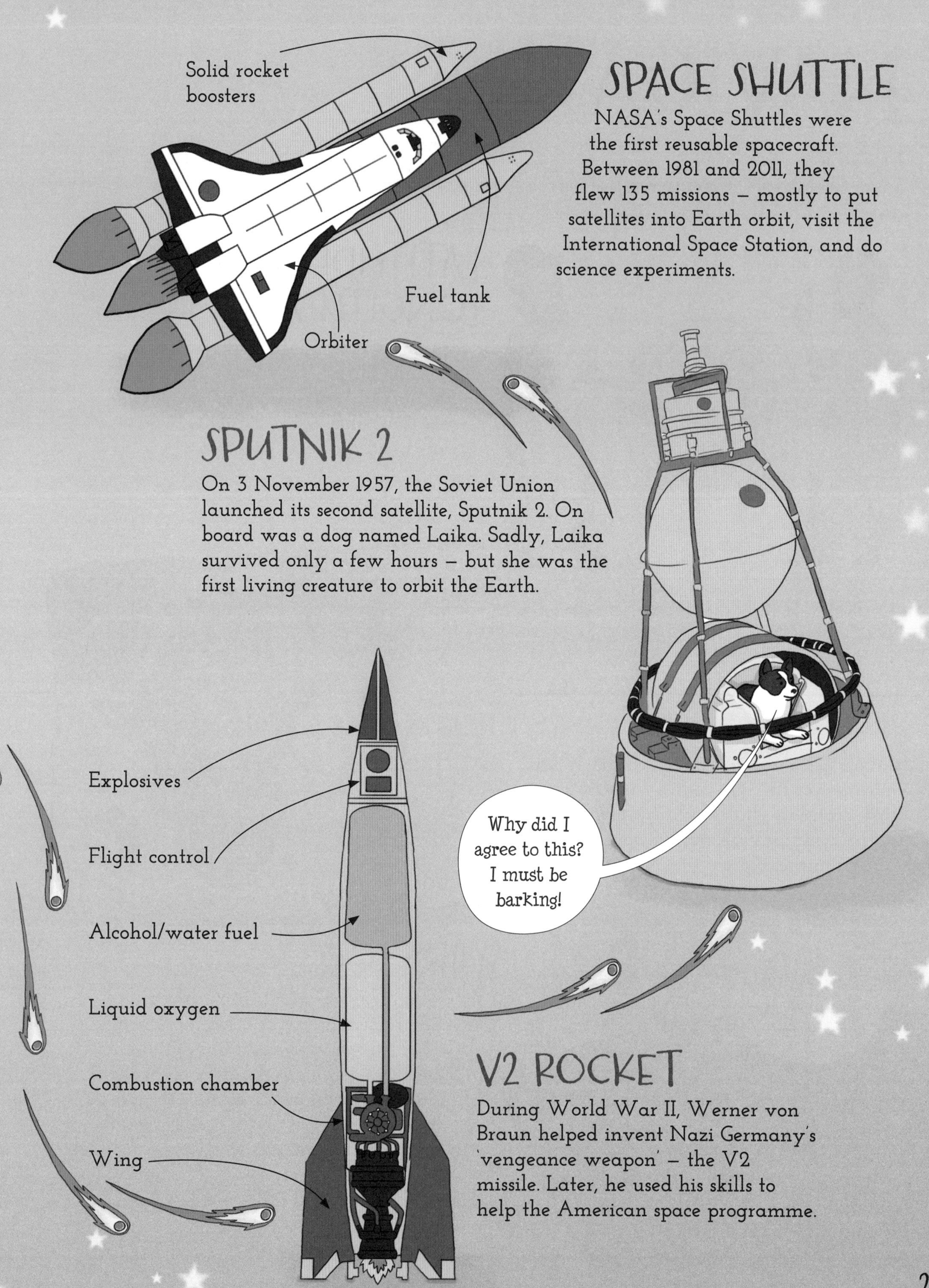

SPACE SHUTTLE

NASA's Space Shuttles were the first reusable spacecraft. Between 1981 and 2011, they flew 135 missions – mostly to put satellites into Earth orbit, visit the International Space Station, and do science experiments.

SPUTNIK 2

On 3 November 1957, the Soviet Union launched its second satellite, Sputnik 2. On board was a dog named Laika. Sadly, Laika survived only a few hours – but she was the first living creature to orbit the Earth.

V2 ROCKET

During World War II, Werner von Braun helped invent Nazi Germany's 'vengeance weapon' – the V2 missile. Later, he used his skills to help the American space programme.

NAVIGATION

Travellers need to know where they are, and where they're going. Finding your way around is called navigating. In early times, this was tricky enough on land, but doubly difficult at sea – at least, until the invention of maps, charts, logs, compasses and other navigational aids.

LATITUDE AND LONGITUDE

Maps have lines of latitude (east-west around the Earth) and longitude (north-south, through the poles). The ancient Greeks Eratosthenes and Hipparchus 'invented' latitude and longitude.

Eratosthenes calculated the circumference of the Earth by measuring shadows cast by the sun. Clever guy!

CHARTS

Sailors' sea maps are called charts. Italian mariners of the 13th century made the first 'portulan' charts, which showed lines of the compass marking out routes for traders and explorers.

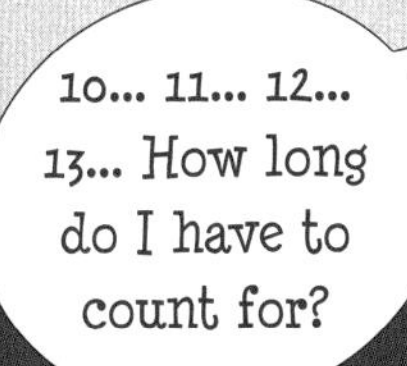

10... 11... 12...
13... How long do I have to count for?

SHIP'S LOG

A ship's log measures speed through water. The earliest one really was a log, to which was tied a rope with knots along its length. Sailors threw the log overboard and counted the knots passing through their hands. That's why a ship's speed is measured in knots.

GPS

These days, mariners have GPS (the Global Positioning System), which uses satellite data to create very accurate electronic charts. The US Department of Defense invented GPS, and 'switched it on' in 1995.

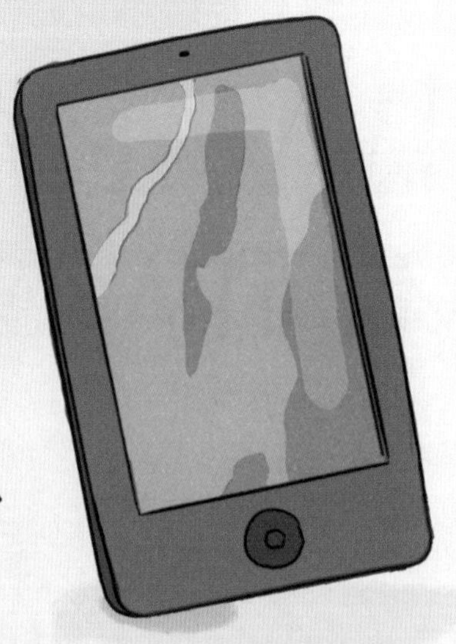

SEA WATCH

If you could keep time at sea, you could work out your position. This H4 'sea watch' of 1761 had taken English clockmaker John Harrison six years to make. Captain James Cook took an H4 watch on his famous voyages to the Pacific.

Harrison's watches were very expensive – roughly one-third of the total cost of fitting out a ship!

QUADRANT

Sailors have long used the stars to navigate by. This 18th-century seaman is using a Davis quadrant to look back at the sun and measure its angle.

FIRST COMPASS

The ancient Chinese invented the compass: a magnetite pointer on a bronze plate. By the 10th century, they'd worked out how to magnetize needles, making better compasses that helped their trading ships navigate at sea.

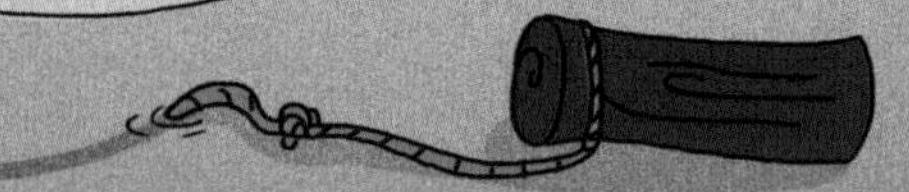

THIS BABY WON'T FLY! (SOME TRANSPORT MISTAKES)

CAR-RRAZY

The first 'car' was a giant, steam-powered three-wheeler designed by French engineer Nicolas-Joseph Cugnot in 1769 – well over a century before the first sensible cars took shape. He also had the world's first car crash when his mad invention smashed into a wall in 1771. Not surprising when you consider it had no brakes...

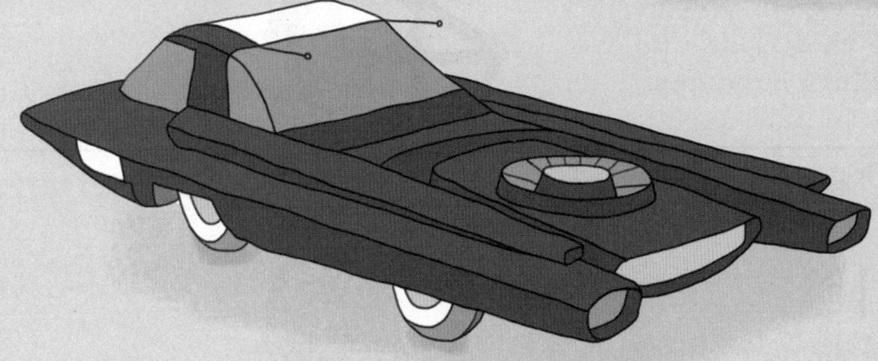

The craziest idea for a car? Hmm... possibly the Ford Nucleon of 1958, with its on-board nuclear reactor. Luckily, it was never built.

IN A FLAP

In the earliest days of manned flight, many pioneers tried to copy the birds. They built ornithopters – flying machines with flapping wings. The incredible thing is, some of them worked...

In 1869, Mr W. E. Quinby got a patent for this flapping wingsuit. Thankfully, he never tried to use it.

In 1928, George White built this foot-powered ornithopter and flew it on a beach in Florida, USA. It managed about 80 wingbeats per minute. Not bad!

MASH-UPS

'Know thyself,' the ancient Greeks warned us. But clearly some people didn't listen. Just look at these mixed-up craft that don't know whether to sink, swim, drive or fly.

In 1936, Soviet engineering student Boris Ushakov designed this flying submarine, supposedly capable of 185 km/h in the air and 5.5 km/h in the water. His bosses finally gave it a test run in 1947. It didn't work.

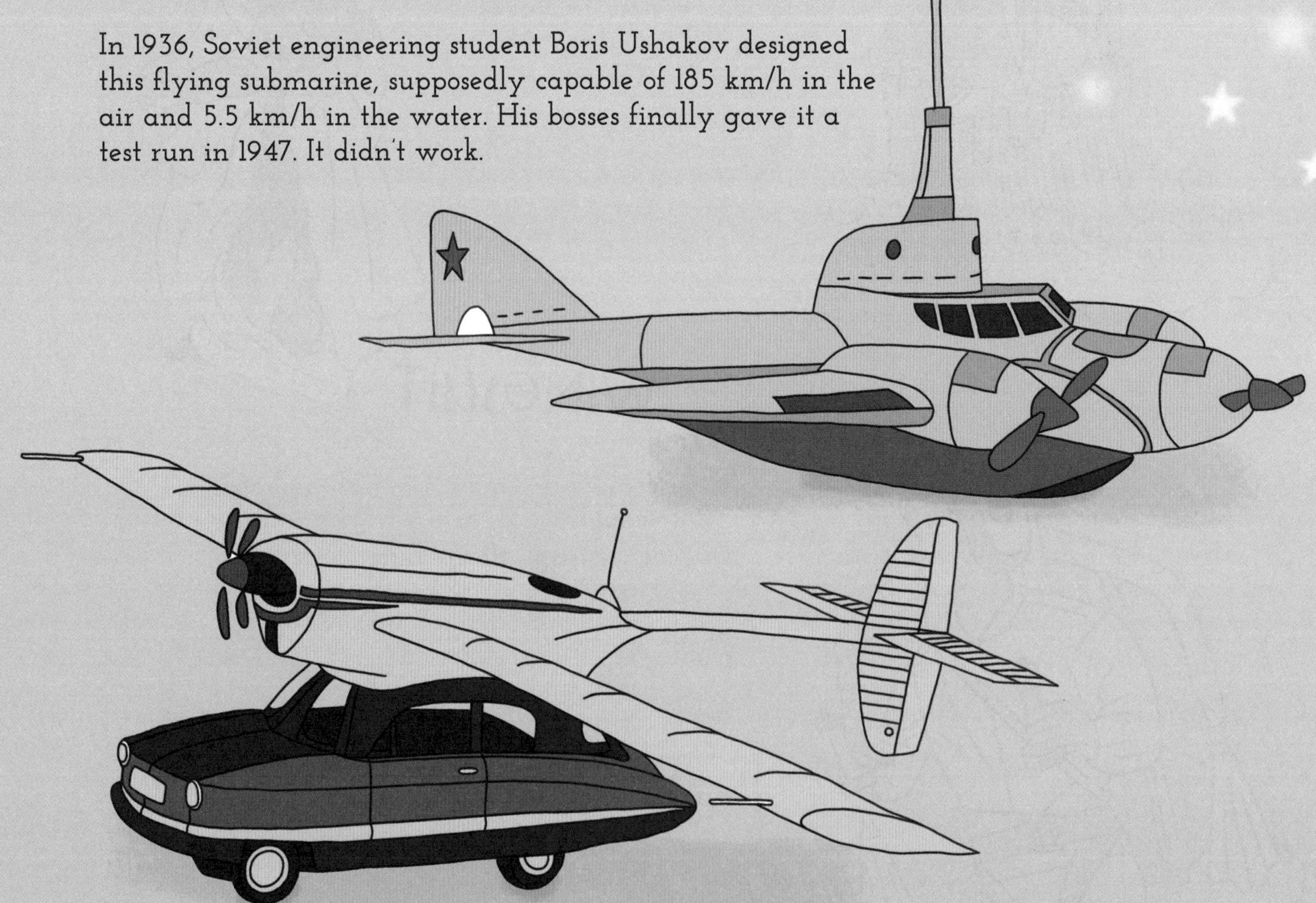

The Convair Model 118 was, as you can see, a car-plane. When the test pilot took it up in November 1947, he forgot to check the fuel, and it ran out of gas, forcing a crash landing. They flew it again in January, but by then Convair had lost interest, and the idea never 'got off the ground'.

WANNA FLY?

Today, daredevils fly solo using hang-gliders, microlights and wingsuits. They make it look easy, even safe… but in days gone by there have been a lot of misguided efforts to invent the perfect personal flight gear.

CHINESE ROCKETS

In Chinese legend, a brave space pioneer named Wan Hu tied rockets to his chair and lit them. One account says he flew, but another says there was a huge explosion: 'When the smoke cleared, Wan and the chair were gone, and were never seen again.'

WINGSUIT

In 1912, tailor-inventor Franz Reichelt created a wingsuit, which he hoped would work like a parachute. To test it, he leapt off the Eiffel Tower in Paris... and fell to his death.

JETPACK

Fancy your own jetpack? Back in the 1950s, the Bell company of America invented this rocket belt – but it flew for only 20 seconds.

This is one VERY small step for mankind.

GLENN MARTIN

Since 1981, New Zealander Glenn Martin has been working on a jetpack. It works... but it costs more than $100,000.

WINGPACK

Swiss pilot Yves Rossy also designs jet 'wingpacks' that fly at up to 300 km/h. In 2008, he flew across the English Channel in just 13 minutes.

HILLER VZ-1

Designed in the mid-1950s for the US Army, the Hiller VZ-1 Pawnee was the brainchild of aircraft engineer Charles Zimmerman. But the Pawnee was small and slow, and its power was puny, so it only rose up to about 10 m.

INDEX

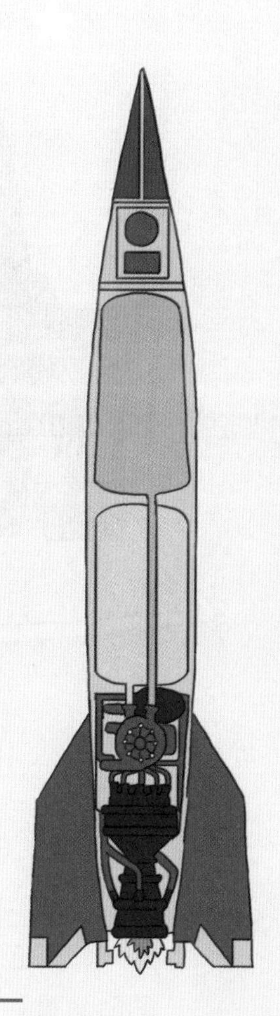

The Author

British-born Matt Turner graduated from Loughborough College of Art in the 1980s, since which he has worked as a picture researcher, editor and writer. He has authored books on diverse topics including natural history, earth sciences and railways, as well as hundreds of articles for encyclopedias and partworks, covering everything from elephants to abstract art. He and his family currently live near Auckland, New Zealand, where he volunteers for the local Coastguard unit and dabbles in art and craft.

The Illustrator

Sarah Conner lives in the lovely English countryside, in a cute cottage with her dogs and a cat. She spends her days sketching and doodling the world around her. She has always been inspired by nature and it influences much of her work. Sarah formerly used pens and paint for her illustration, but in recent years has transferred her styles to the computer as it better suits today's industry. However, she still likes to get her watercolours out from time to time, and paint the flowers in her garden!

INCREDIBLE INVENTIONS ON THE MOVE

Written by Matt Turner
Illustrated by Sarah Conner

外语教学与研究出版社
FOREIGN LANGUAGE TEACHING AND RESEARCH PRESS
北京 BEIJING

奇思妙想发明史

神奇的光

马特·特纳（英）著
萨拉·康纳（英）绘
刘姣 译

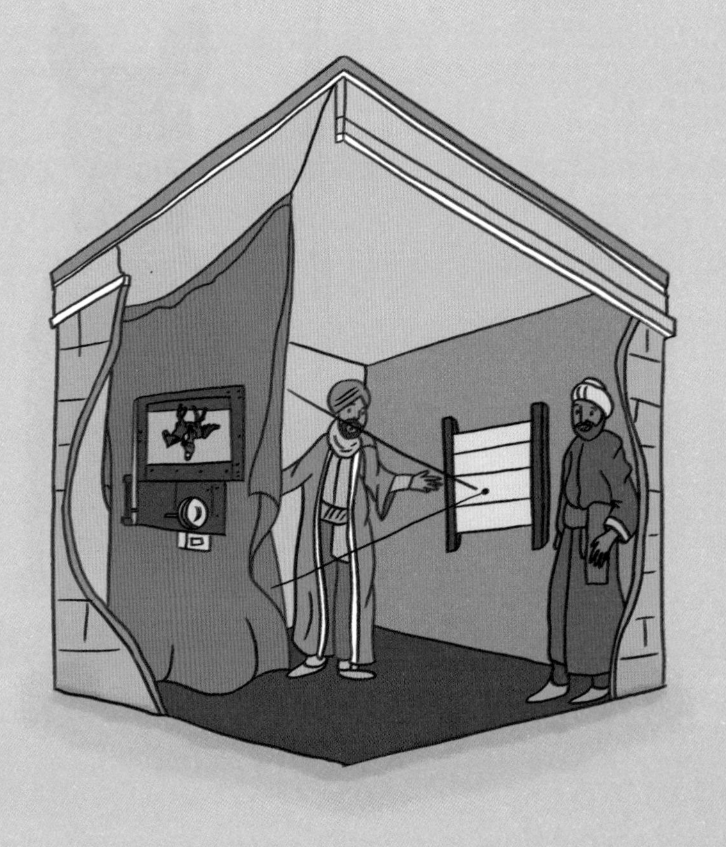

外语教学与研究出版社
FOREIGN LANGUAGE TEACHING AND RESEARCH PRESS
北京 BEIJING

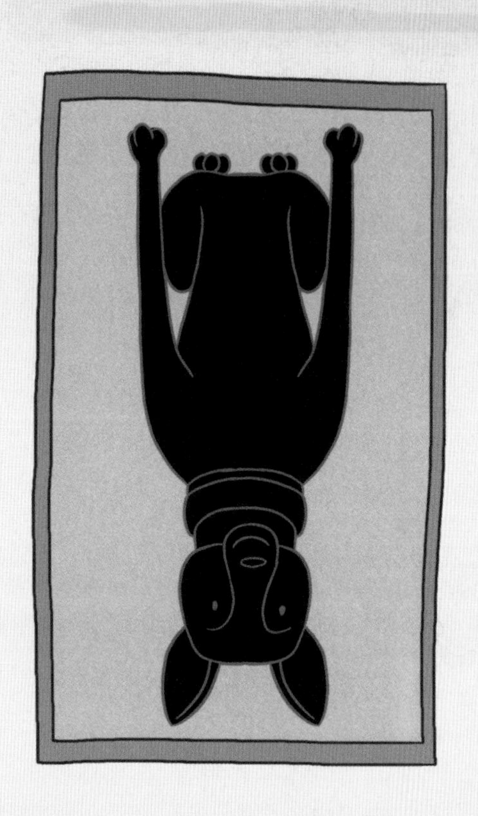

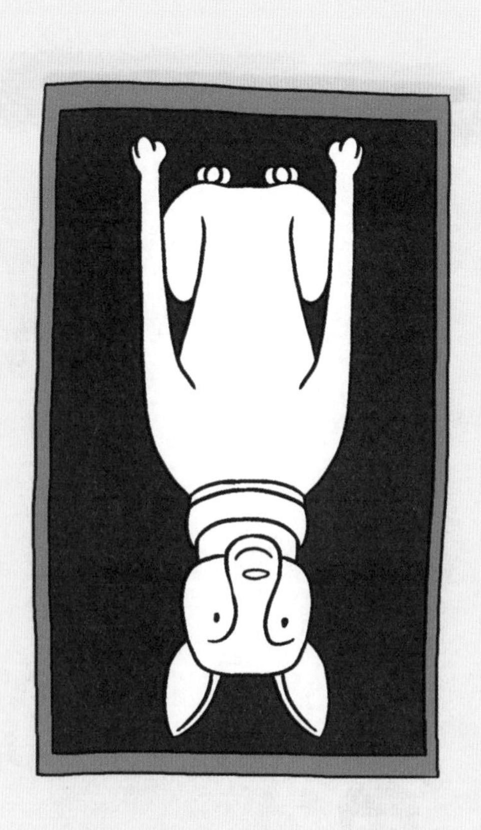

在发明放大镜前，人们如何把物体放大了来观察呢？当然是通过装水的玻璃碗！

后来人们又发明了显微镜和望远镜……让我们一起来回顾人类历史上的一系列发明创造吧。有了它们，我们得以观察万物——从恒星到细胞，甚至肉眼完全不可见的事物。在这本书里，你会发现：

- 托马斯·爱迪生发明出“对的”灯泡前，试验了3,000种不同的灯泡
- 最原始的照相机就是一间黑屋子（墙上有个洞）
- 人们观看西尼拉玛全景电影时，会激动地尖叫
- 磁共振成像扫描仪能“看见”肿瘤
- 激光可以切割钢材

还有更多的发现等着你。不过也有一些脑洞大开的发明（比如给马戴的眼镜，为什么不可以？），以及一些离谱的预测——有人认为永远不会有万家灯火的那一天……

伽利略（1564—1642）用自制望远镜观察近处和远处的物体，观察对象小到昆虫，大到行星。

目录

导语：走进光的世界

我们的祖先首次学会生火和控制火时，人类的命运就被改写了。他们学会了制造光（和热量）后，就再也不用日落而息。晚上还可以生火取暖、烹煮食物，再也不用吃生肉了！

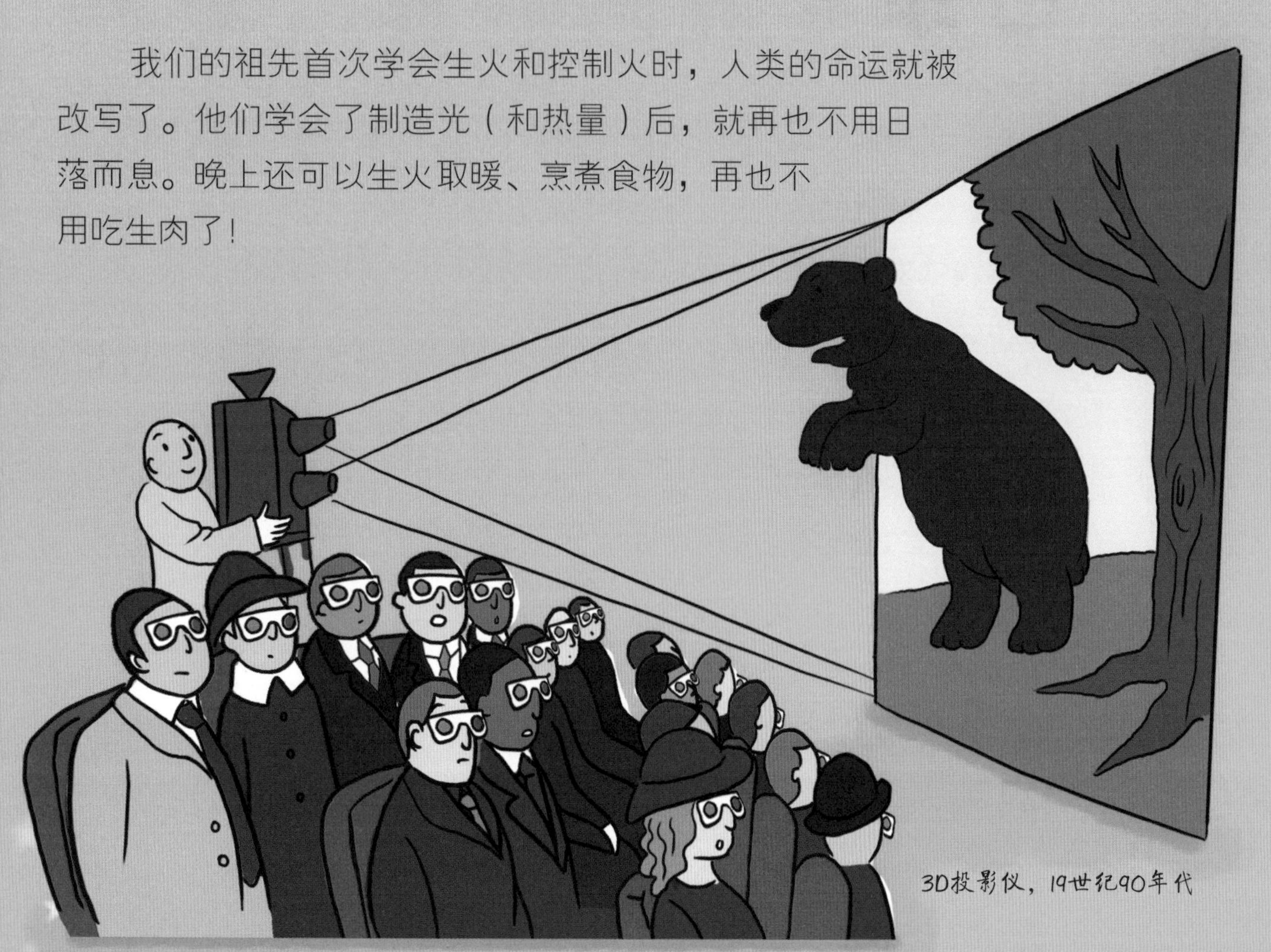

3D投影仪，19世纪90年代

吃得更好甚至会使我们的大脑变大，让我们更聪明。早期的人类还利用火驱赶讨厌的虫子和野兽，利用火烧空树干做成独木舟航海，寻找新的陆地。当然，这些进步发生在很久以前，而且我们无法确认发明者是

至少2,000年前的有放大功能的玻璃碗

谁。更确切地说，这些进步是人类社会的“巨大飞跃”。

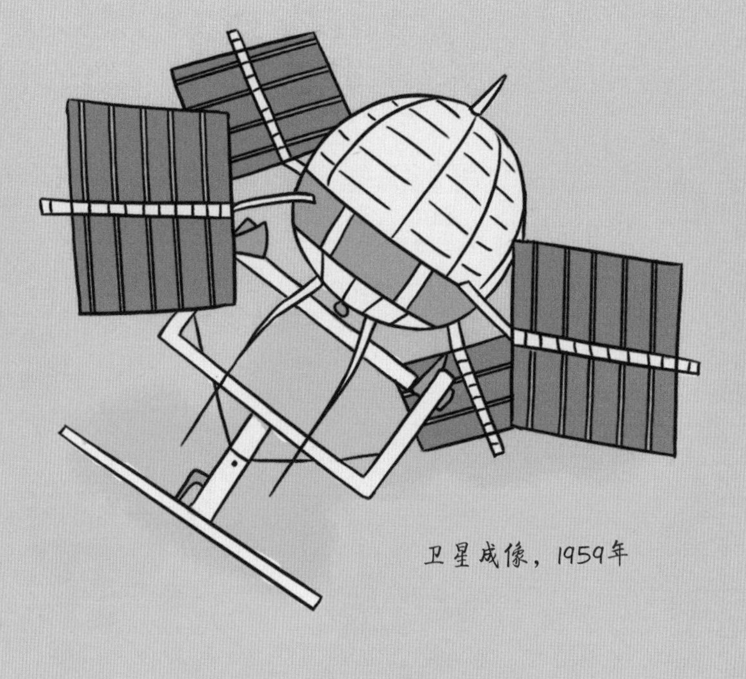

卫星成像，1959年

除了制造光，人类还学会了用镜片聚光。人类首次擦亮用水晶做的镜片，或者将玻璃球装满水使用时，发现这些简单的镜片可以“弯曲”光线，将物体放大。在显微镜、望远镜、眼镜和电影摄像机里，光都扮演着至关重要的角色。

加入我们的历史环游之旅吧，了解从早期的钻木取火到现代激光和卫星摄影的发展历程。我们将探索更为新奇的观察事物的方式，比如无线电波、微波、X射线和声波的使用，它们的存在给我们带来了雷达和医用扫描仪。

我们还会趣谈一些关于光的异想天开的发明，比如给鸡和马戴的眼镜！

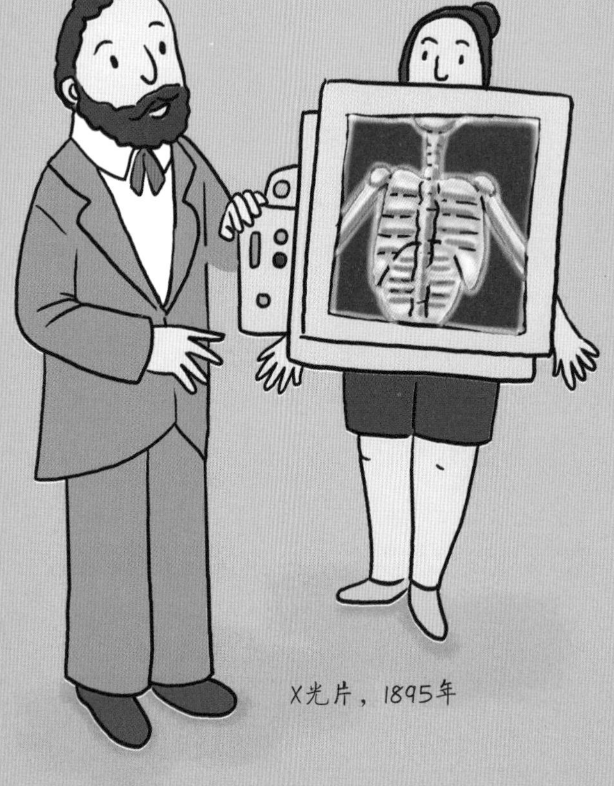

X光片，1895年

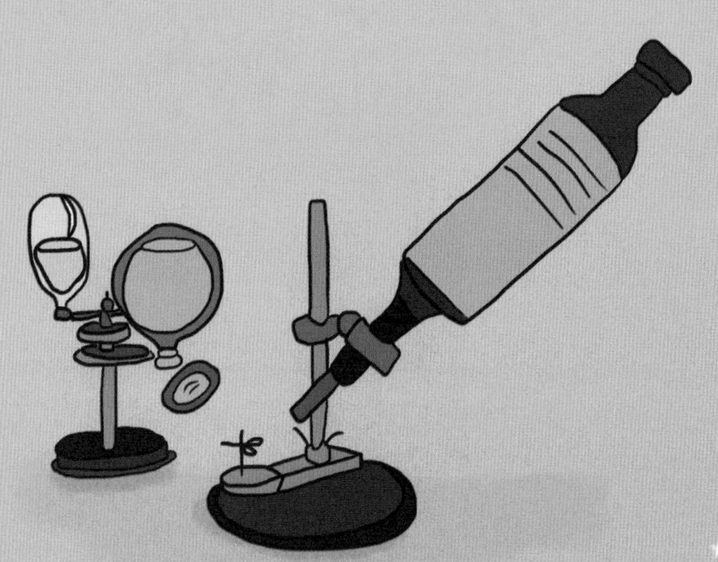

胡克的显微镜，1665年

人类如何制造光？

你现在很可能开着电灯，对吧？是用开关开的吗？我们的祖先可没这么幸福，为了制造光，他们用尽了各种办法，最开始是钻木取火，后来又制作了火柴或者巨大的电池。

火

大约100万年前，我们的祖先学会了钻木取火，通过摩擦产生热量。他们还通过击打坚硬的岩石产生火花。焚烧过的骨头残留物显示，他们曾用火烹煮过肉食。

油灯

油灯的历史可以追溯到12,000多年前，那时油灯一般是用贝壳或雕刻过的石头做的，或者就是一个黏土杯，用动物脂肪当燃油。罗马油灯烧的是橄榄油，有的灯每盏有10或12根灯芯。

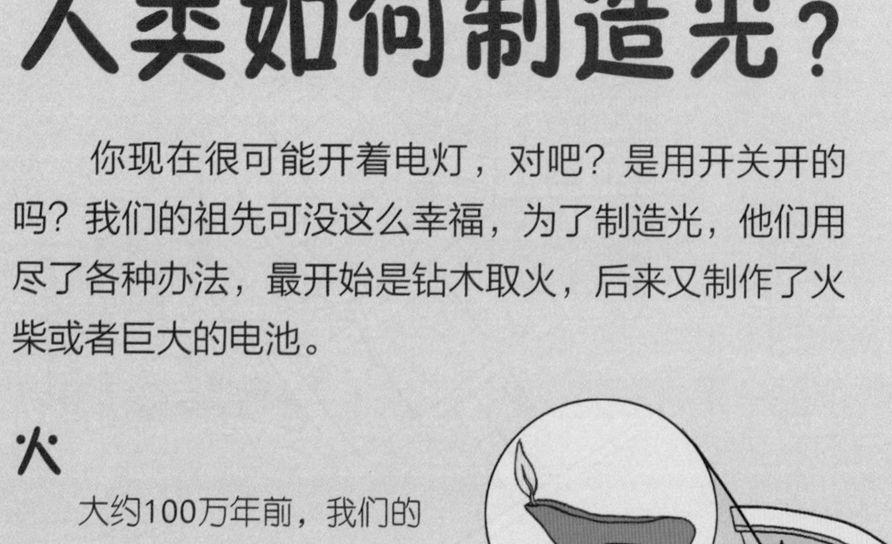

灯塔

位于埃及亚历山大城的法罗斯岛灯塔建于2,000多年前，高度将近135米，数个世纪后，在一场地震中倒塌。这座灯塔的光是由顶端的火炉产生的。后来的灯塔都用上了电灯泡，这多亏了爱迪生的好点子（见右页）。

火柴

大约1,000年前，中国人发明了火柴，当时叫做“火寸条”。19世纪20年代，质量可靠、通过摩擦砂纸点火的火柴首先出现在英国。这种早期的火柴含有磷，导致火柴厂的工人和卖火柴的商贩患上重病，比如右图这个男孩。

白炽光

1801年，英国科学家汉弗莱·戴维发现了白炽光（即通过热量产生的光）。为了把一根铂丝加热至发光，他用了当时世界上最强的电池——2,000个相连的电池。厉害！

喂，你这种电池有袖珍手电筒大小的吗？

戴维的安全灯

1815年，戴维因发明了给煤矿工人用的“安全灯”而出名。这种灯的火焰被完全罩起来了，以减少煤气爆炸的风险。但不是很好用：发出的光很昏暗，而且煤气爆炸依旧会发生。唉！

爱迪生的好点子

第一个电灯泡并不是美国人托马斯·爱迪生发明的。但在试验了3,000多种灯泡和6,000余种不同的灯丝后，爱迪生于1879年发明了第一个质量可靠的灯泡！（灯丝就是灯泡里那根发光的细丝。）

显微镜的发明

放大镜、眼镜、隐形眼镜和显微镜所依据的原理如下：一片曲面玻璃，也就是镜片，能聚焦光线，让物体离我们的距离显得比真实距离更近。从古至今，镜片给我们带来了很多科学发现（也使我们在公交车上多丢一样东西——眼镜）。

胡萝卜、青蛙腿、牛奶……

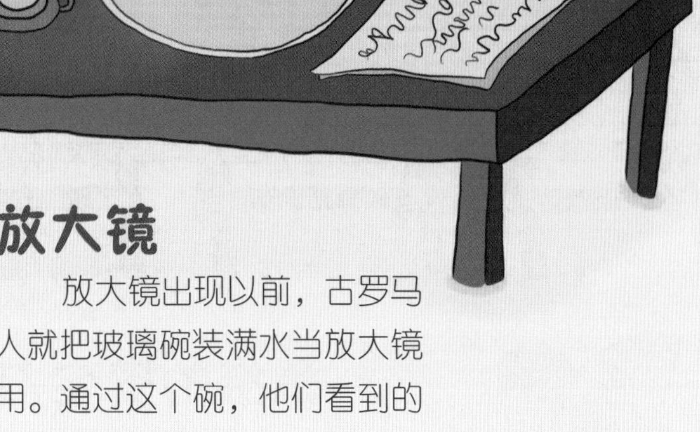

放大镜

放大镜出现以前，古罗马人就把玻璃碗装满水当放大镜用。通过这个碗，他们看到的东西会被放大。

眼镜

阿拉伯人很久以前就懂得光学原理，但在西方，英格兰修道士罗杰·培根在1268年首次撰写了关于镜片的著作。在此后20年的时间里，意大利人发明了“夹式”眼镜。

真希望我看不见这些青蛙腿。

远视近视两用眼镜

美国政治家本杰明·富兰克林（1706—1790）发明了远视近视两用眼镜，即一副装有两种不同焦距镜片的眼镜。有次去法国，富兰克林戴了这种眼镜来看自己吃的东西和同桌的用餐者。

隐形眼镜

1888年，德国医生阿道夫·菲克设计了最早的实用隐形眼镜。他先用兔子做了实验。（但他怎么知道这种隐形眼镜是否有效呢？）

我现在拿着几根胡萝卜？

电子显微镜

20世纪30年代，德国电视工程师马克斯·克诺尔发明了性能超强的扫描电子显微镜（SEM）。扫描电子显微镜首先被用于近距离观察金属，还能给虫子拍照片，看得人头皮发麻……

胡克的显微镜

英格兰人罗伯特·胡克最早使用cell（细胞）这个词来描述生命的基本单位。1665年，他通过显微镜观察到了细胞。

观察肉眼不可见的物体

显微镜通过非常强大的镜片来放大极其微小的东西。最早的显微镜是复式显微镜，即几根装了几套镜片的管子，通过共同作用来放大物体。使用这种显微镜，科学家首次观察到了细菌、血细胞和酵母。这种显微镜（下图是俯视图，左图是侧视图）是荷兰人安东尼·范列文虎克（1632—1723）发明的。你可以把标本固定到标本针的针尖上，通过薄板上放大率为275倍的镜片进行观察。

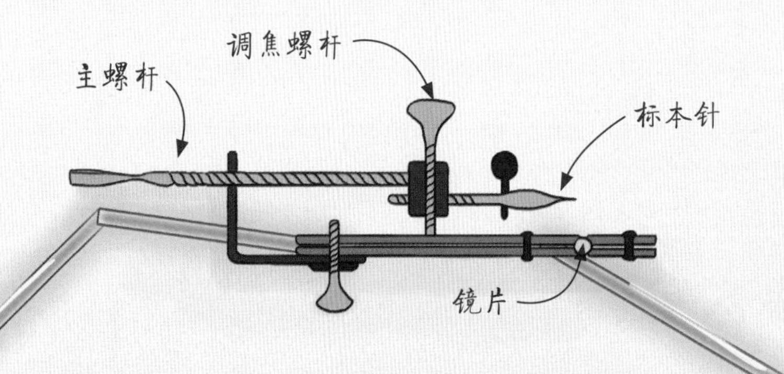

望远镜的发明

有了望远镜，天文学家就可以深入观察太空，揭开行星、卫星和恒星的奥秘。早期的望远镜使用了镜片，很像显微镜，但现在我们也用射电望远镜来探索宇宙。

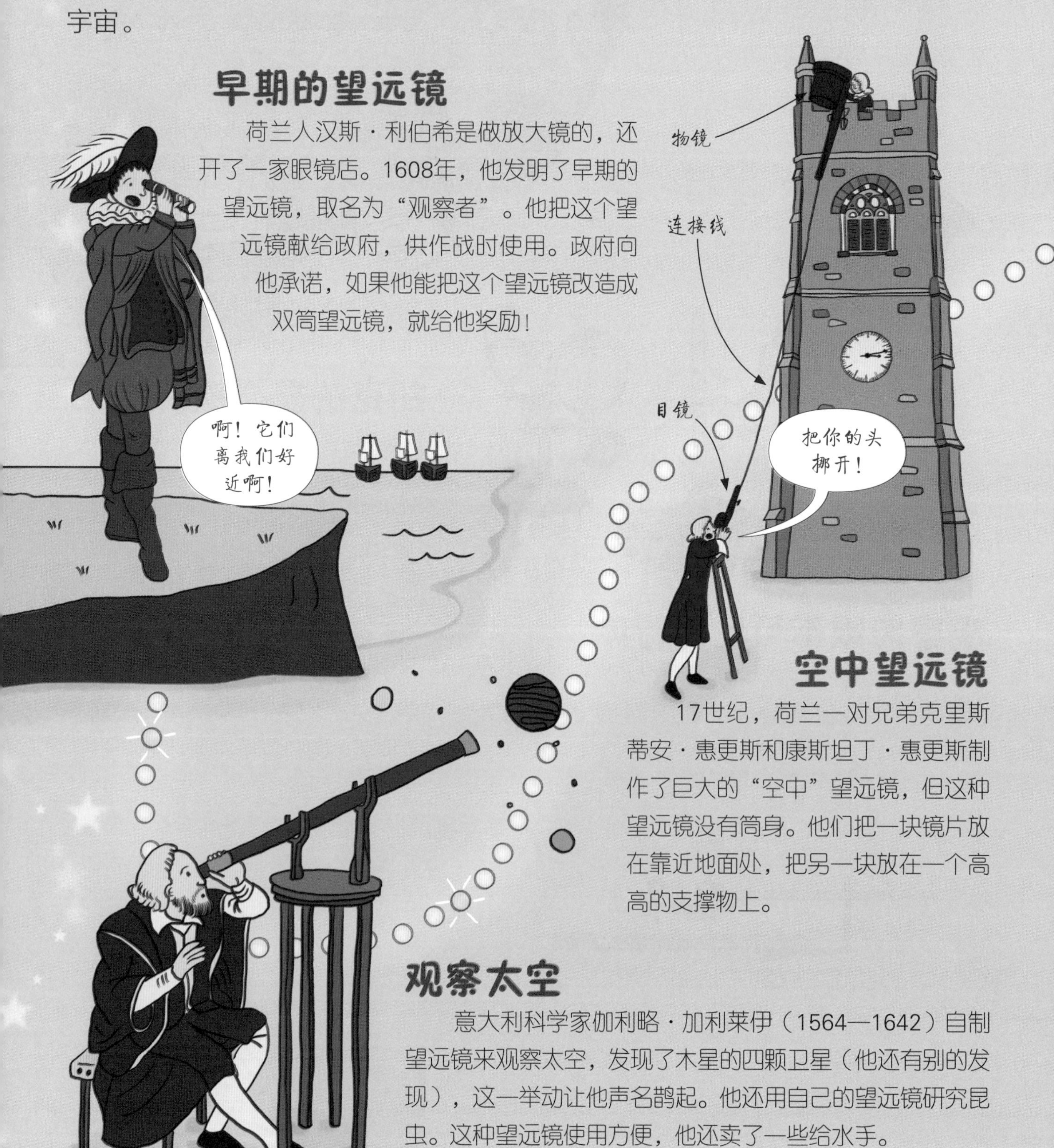
啊！它们离我们好近啊！
物镜
连接线
目镜
把你的头挪开！

早期的望远镜

荷兰人汉斯·利伯希是做放大镜的，还开了一家眼镜店。1608年，他发明了早期的望远镜，取名为“观察者”。他把这个望远镜献给政府，供作战时使用。政府向他承诺，如果他能把这个望远镜改造成双筒望远镜，就给他奖励！

空中望远镜

17世纪，荷兰一对兄弟克里斯蒂安·惠更斯和康斯坦丁·惠更斯制作了巨大的“空中”望远镜，但这种望远镜没有筒身。他们把一块镜片放在靠近地面处，把另一块放在一个高高的支撑物上。

观察太空

意大利科学家伽利略·加利莱伊（1564—1642）自制望远镜来观察太空，发现了木星的四颗卫星（他还有别的发现），这一举动让他声名鹊起。他还用自己的望远镜研究昆虫。这种望远镜使用方便，他还卖了一些给水手。

寻找土星的卫星

德裔英国人威廉·赫舍尔是一位天文学家，他在1789年建造了下图这个长达12米的巨型望远镜。这是个反射式望远镜，与牛顿的望远镜相似，但赫舍尔改进了内置的镜子装置。他用这个望远镜观察土星的卫星。其中一颗卫星叫做土卫一，看起来像死亡之星，非常恐怖。

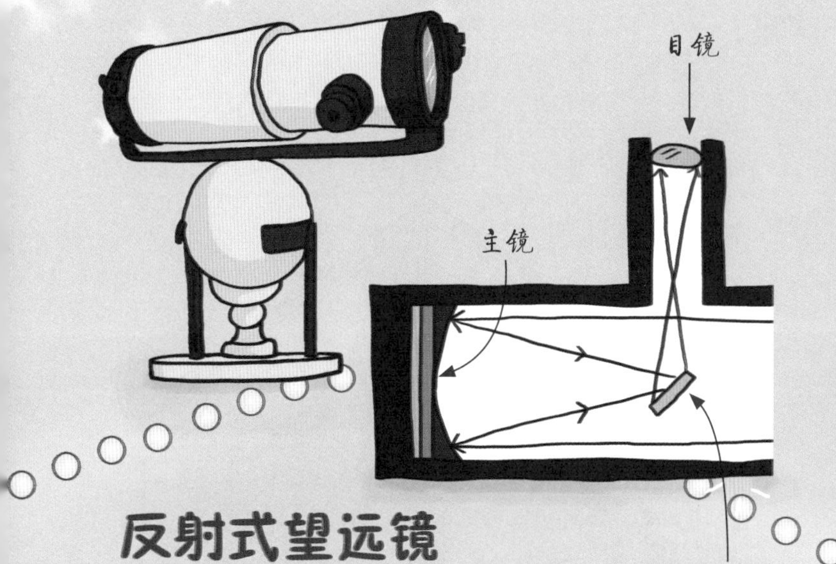

反射式望远镜

早期的望远镜镜片容易把光线分离成彩虹的颜色，使图像模糊不清。1668年，英国科学家艾萨克·牛顿用反射式望远镜解决了这个问题。这种望远镜用镜子取代镜片来捕捉图像。

射电望远镜

现代射电望远镜不用镜片来“看”，而像电视天线一样“听”。1931年，美国人卡尔·央斯基在一片土豆田里建造了第一架射电望远镜。他把天线连到无线电接收器上，来监测来自太空的无线电波（即星球活动产生的能量），听到了平稳的嘶嘶声。

射电天文学

央斯基的发现给了他的美国同胞格罗特·雷伯灵感。1937年，雷伯在自家后花园里建造了一架射电望远镜，用来制作一张太空无线电信号“地图”。他是当时全世界唯一一位射电天文学家！而今天，全世界有成千上万名射电天文学家……

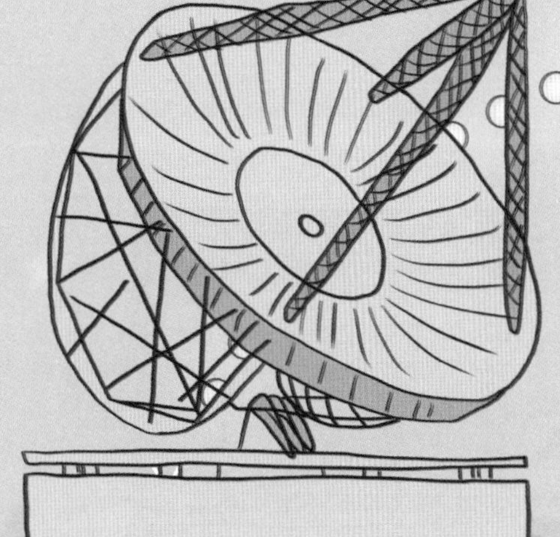

照相机发展史

现代照相机非常复杂，让人觉得不可思议，但基本原理来自“暗箱”（意思是“黑屋子”），或称针孔照相机。下图是一个箱子，光线只能从其中一面墙上的小孔射入。光线进入后，像投影仪一样在对面的墙上投射出图像（图像上下颠倒）。如果在那面墙上贴上感光胶片或者感光纸，就能捕捉这个图像作为照片。

暗箱

阿拉伯科学家海桑（965—1039）描述了针孔照相机的工作原理，天文学家和艺术家均采用了他的理论。

早期的摄影术

摄影术的真正发明者是法国人尼塞福尔·涅普斯和路易·达盖尔。尼塞福尔试验了感光化学物质（氯化银、沥青和薰衣草油），并在1816年左右拍出初级的照片。1839年，达盖尔研发了“达盖尔式”摄影法。他将一张抛过光、表面涂有化学物质的银片放在暗箱里曝光。没错，这种相机就叫“达盖尔”（下图），能拍出非常清晰的照片。

感光纸照片

1800年左右，英国科学家托马斯·韦奇伍德用感光纸做了一个实验。他把感光纸放在有阳光照射的窗台上，拍下了树叶和其他物体的“阴影照片”。

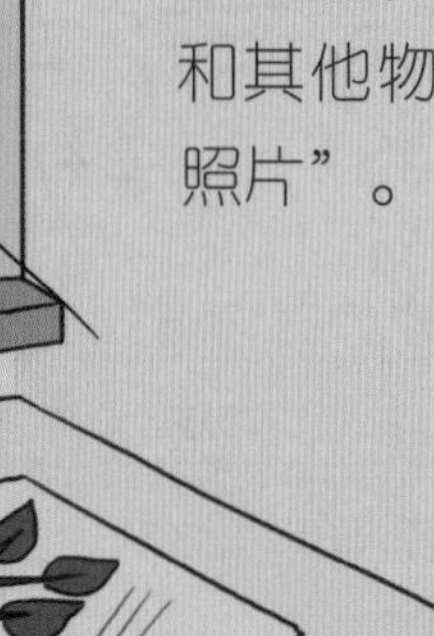

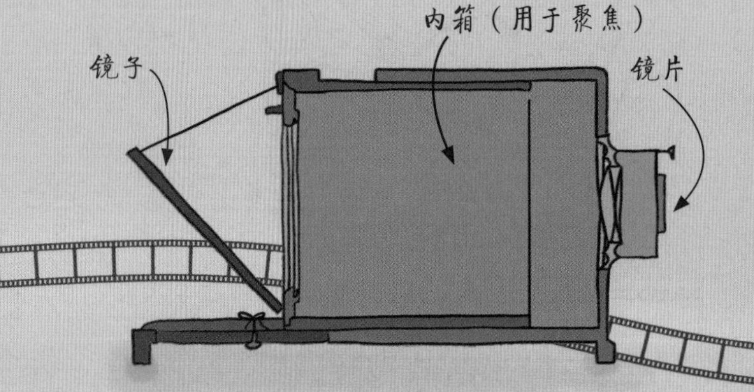

宝丽来照片

用宝丽来照相机拍的照片可以立马打印出来，这种照相机是1947年美国人埃德温·兰德发明的。

数码照相机

1969年，威拉德·史密斯和比尔·博伊尔发明了电荷耦合器件（CCD）这种电子小装置。他们的发明终结了胶片照相机的时代，迎来了数码照相机的黎明。1975年，斯蒂芬·萨松发明了数码照相机。

伊斯门的柯达照相机

1884年，美国人乔治·伊斯门发明了胶片，引发了摄影术的重大革命。胶片是柯达照相机的特色，他在1888年为柯达照相机申请了专利。小而便宜的照相机的问世，比如布朗尼相机，意味着从此人人都可以成为摄影师。

正像与负像

同样是在19世纪30年代，英国科学家威廉·福克斯·塔尔博特发明了光力摄影法，比达盖尔摄影法更先进，也更快捷。有了光力摄影法，你就可以制作一张负像（前后颠倒的图像），然后用它制作正像。

五花八门的勘测手段

勘测的意思是丈量土地。勘测非常重要，因为有了勘测才有地图和全球定位系统（这样爸妈开车时就不会迷路了）。勘测人员勘测时还能去很多刺激和危险的地方。

我确定它就在附近某个地方……

勘测山岭

古埃及人将长绳打结来丈量距离。中国古人（下图）在一根杆子上放一个量角器，通过测量角度来计算山的高度。几个世纪以来，这种测量方式以及与其类似的方法被人们广泛使用。

最早的地图

巴比伦世界地图（收藏于现在的伊拉克）是已知的世界上最早的地图。大约2,500年前，一位无名人士把这幅地图刻在一块泥板上，地图上的圆环代表围绕陆地的海洋。

卫星成像

现在，绕地球运行的卫星能拍出非常清晰的照片，你在花园里晒太阳都能被拍到。发射于1959年的探险者6号卫星拍摄了一些最早的太空照片，照片非常模糊，但标志着卫星成像时代的到来。

经纬仪

过去，勘测人员为了绘制新帝国的地图，会带着经纬仪出门。经纬仪实际上就是能精确测量角度的望远镜。有些经纬仪个头很大，比如左下图这个就重约100千克，是杰西·拉姆斯登于18世纪80年代在英格兰制作的。美国独立之前，年轻的乔治·华盛顿（美国首任总统，译者注）是北弗吉尼亚的第一个勘测员……

航空摄影术

1858年，法国人纳达尔发明了航空摄影术。右图是他做的最大的一个气球，叫做“巨人”。纳达尔还发明了航空邮件！1870至1871年间，他用一队气球将200多万封信件偷运出巴黎，当时的巴黎正被德国人四面包围。

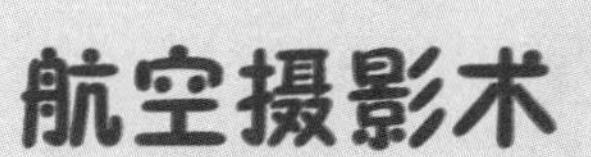

火箭照相机

1897年，瑞典天才阿尔弗雷德·诺贝尔（因诺贝尔奖出名）开创了火箭航空摄影术。1906年，德国工程师阿尔贝特·毛尔（右图）设计了更有效的摄影火箭，这种火箭以压缩空气为动力。

军事侦察

毛尔发明的“摄影火箭”从未发射过，因为航空摄影师很快用上了飞机。1903年，飞行先锋威尔伯·莱特首次飞行成功。第一次世界大战期间，作战双方都使用了飞机来记录彼此的位置。

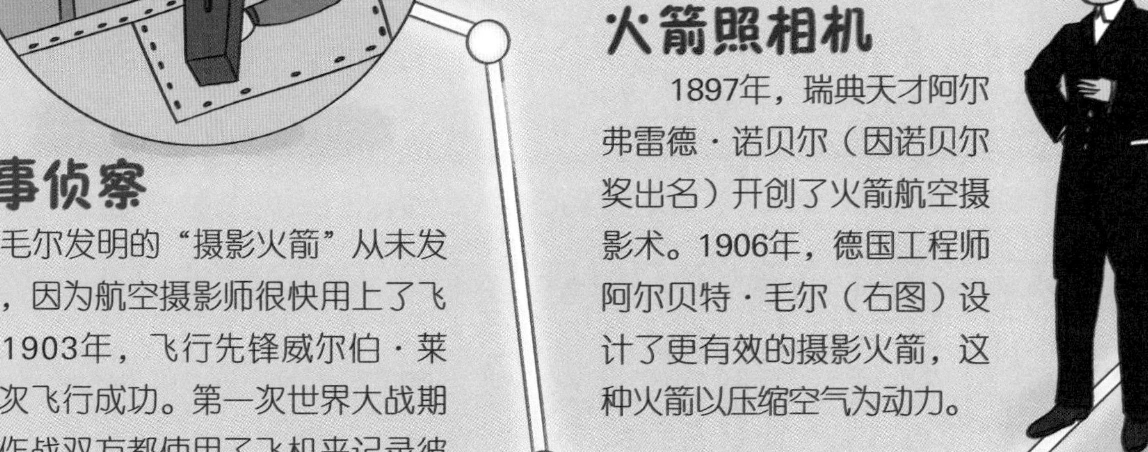

电影是怎么来的？

有谁会不喜欢看电影呢？如果有，那就是你的老祖先们。毕竟，电影摄影机最早出现于19世纪90年代，历史还很短。但很久以前，发明家们就开始制作一些巧妙的玩具来迷惑我们的眼睛，让我们相信自己看到的是会动的图像——电影。

转盘活动影像镜

其中一种玩具就是比利时人约瑟夫·普拉托在1832年发明的转盘活动影像镜。这个影像镜有两个圆盘，其中一个在边缘处绘有连续的图像。当圆盘转动时，你通过一个狭孔看一面镜子，就能看到图像“连在一起”，成了动画。

动物实验镜

与转盘活动影像镜相似的玩具还有动物实验镜（1879年由埃德沃德·迈布里奇发明），这个实验镜首次产生了活动影像。

第一台电影摄影机

有人称托马斯·爱迪生发明了电影摄影机，但真正的发明者是法国人路易·勒普兰斯。1888年，他用这台摄影机（右图）在英格兰利兹拍摄了一些影像。但两年后他就失踪了，所以他的发明从没给他带来任何财富。

活动物体连续摄影机

爱迪生和他的助手比尔·迪克森发明了一台电影摄影机，叫做活动物体连续摄影机。1891年，爱迪生向公众展示了他的电影放映机。这台机器可以放电影，但每次只能供一个人观看。

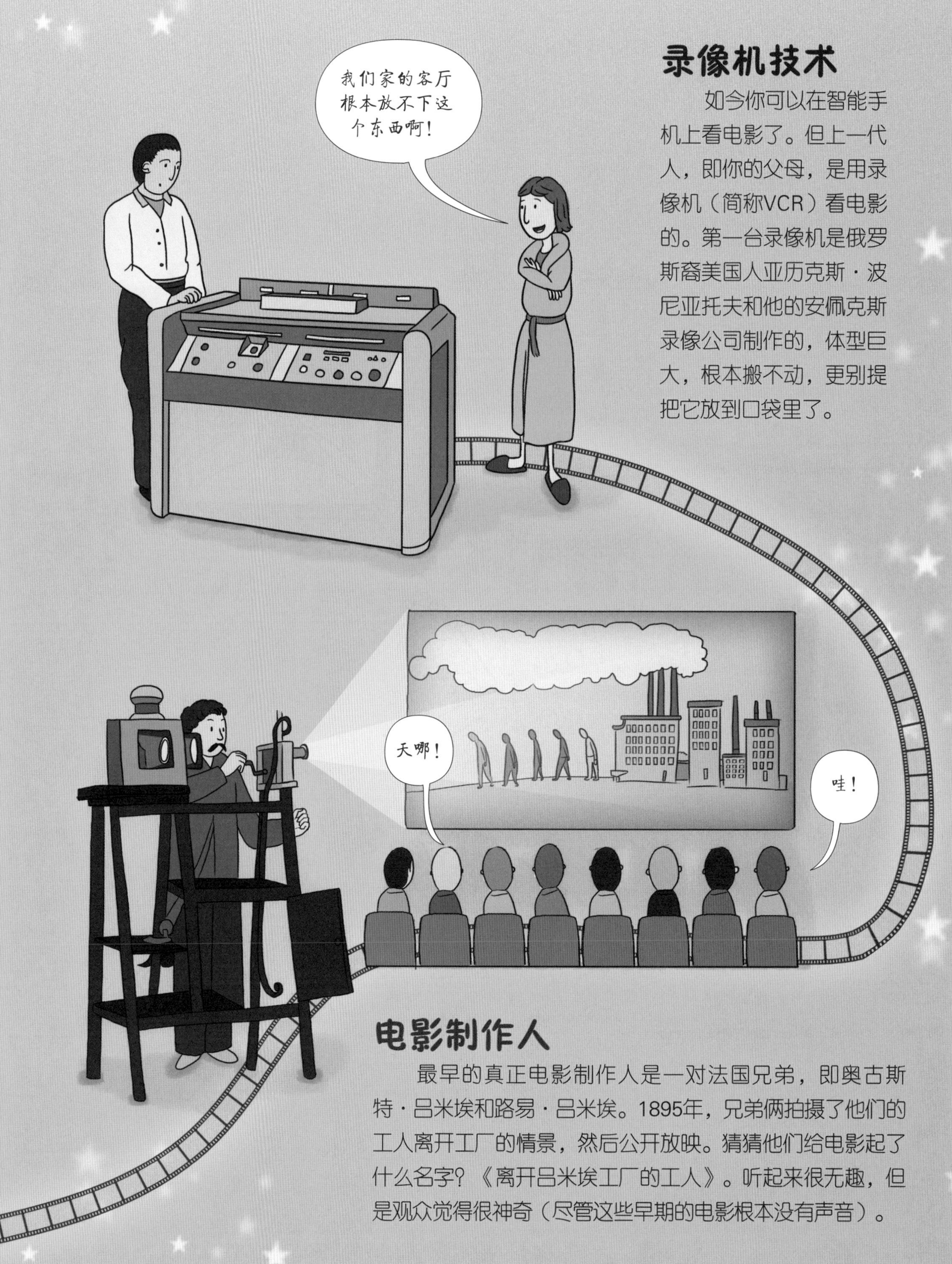

录像机技术

如今你可以在智能手机上看电影了。但上一代人，即你的父母，是用录像机（简称VCR）看电影的。第一台录像机是俄罗斯裔美国人亚历克斯·波尼亚托夫和他的安佩克斯录像公司制作的，体型巨大，根本搬不动，更别提把它放到口袋里了。

电影制作人

最早的真正电影制作人是一对法国兄弟，即奥古斯特·吕米埃和路易·吕米埃。1895年，兄弟俩拍摄了他们的工人离开工厂的情景，然后公开放映。猜猜他们给电影起了什么名字？《离开吕米埃工厂的工人》。听起来很无趣，但是观众觉得很神奇（尽管这些早期的电影根本没有声音）。

3D技术

3D电影很有趣，但就像古老的转盘活动影像镜一样，只是欺骗大脑的巧妙把戏。你或许以为，制作和观看3D电影是现代才有的事，其实它已经有150多年的历史了，比电影摄影机的历史还要长。

立体镜

3D电影问世很早，其历史几乎可以追溯到摄影术发明之时。1849年发明的布鲁斯特立体镜内装有两张照片，一张在左，一张在右，各是从一个位置稍有变化的地方拍摄的。两张照片放在一起观看，就会产生3D效果。

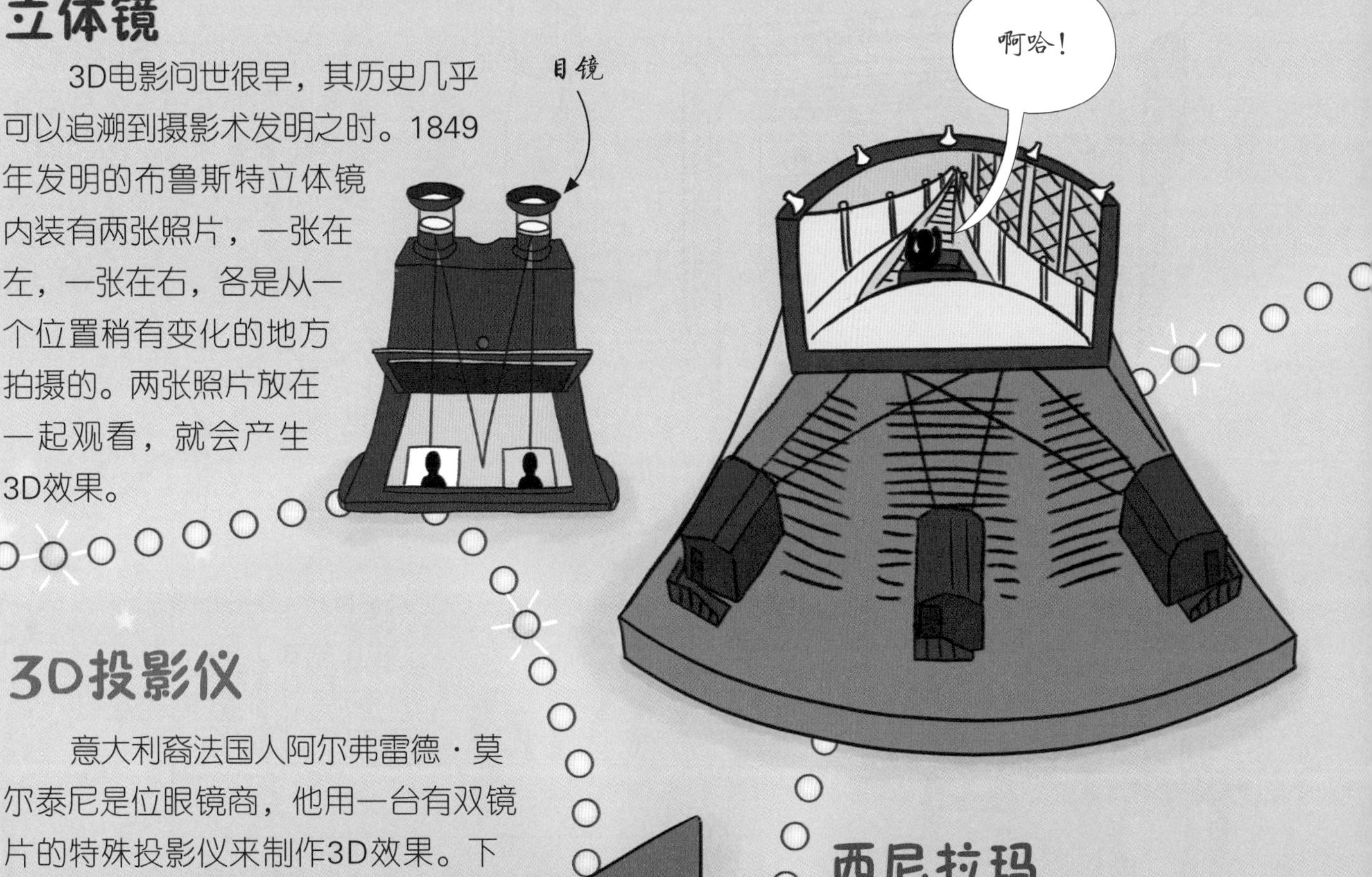

3D投影仪

意大利裔法国人阿尔弗雷德·莫尔泰尼是位眼镜商，他用一台有双镜片的特殊投影仪来制作3D效果。下图的观众正在观看他制作的关于大自然的幻灯片。

西尼拉玛全景电影技术

比莫尔泰尼的设备更先进的是西尼拉玛全景电影技术，由美国人弗雷德·沃勒发明。这一技术使用了巨大而有曲度的屏幕，用三台电影放映机产生了“连在一起”的画面。1952年首次放映时，观众非常兴奋，连连尖叫！但西尼拉玛全景电影技术设计极其复杂且十分昂贵，直到人们找到了只用一台放映机的方法。

3D扫描

1859年，在巴黎弗朗索瓦·威廉的圆形工作室里，安装在墙上的24台照相机可以给人拍照。威廉用一个投影仪将每张照片的轮廓复印到一块黏土上，然后把它雕刻成3D“照相雕塑”。3D扫描和印刷就这样诞生了。

特效

比威廉的设备更现代的技术叫做“摄影测量法”，被应用于电影业。具体方法是先用数码相机给演员拍多张照片，然后动画师用这些照片来制作你在电影上看到的特效。

可以扫描的应用程序

2009年，美国软件设计公司欧特克开发出了123D Catch软件（一种3D建模工具）。你可以用这种软件给物体拍摄3D照片，然后在电脑上处理，甚至以3D形式打印出来。如果弗朗索瓦·威廉那个年代有这个软件，那它一定会受到他的青睐！

人体扫描技术

到现在为止，我们已经探索了细菌、外太空、地球……现在来看看人体内部吧。听起来恶心？也许吧，但这挽救了无数人的生命。今天，这种技术叫做放射造影术或放射学，因为它使用了能穿透我们身体的辐射波。放射治疗师还能用声波给我们的内脏拍照。

早期的内窥镜

大约在1800年，德国医生菲利普·博齐尼发明了“光导镜”。它由一根蜡烛和一只镜片构成，用于观察耳朵或喉咙内部。这是最早的内窥镜之一。

给士兵拍摄X光片

不久，科学家便意识到X射线可以挽救生命。波兰裔法国化学家玛丽·居里是杰出的放射线研究者。第一次世界大战期间，她派专门的医用卡车前往战场为受伤士兵拍摄X光片。在放射性研究方面取得的成就为居里夫人赢得了很多奖项。

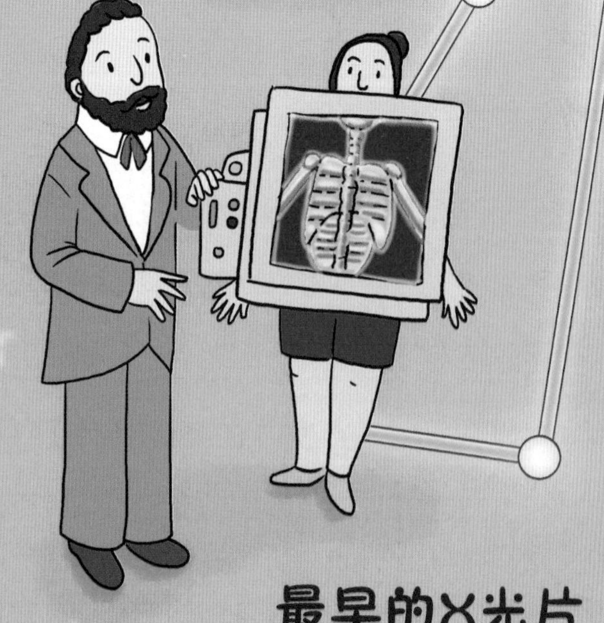

最早的X光片

1895年，威廉·伦琴在德国发现，某些射线可以直接穿过硬纸板，投射到屏幕上发光。他拍摄了第一张X光片，照片显示了他的妻子安娜的手骨。于是人们意识到可以借此探索人类身体的内部构造。

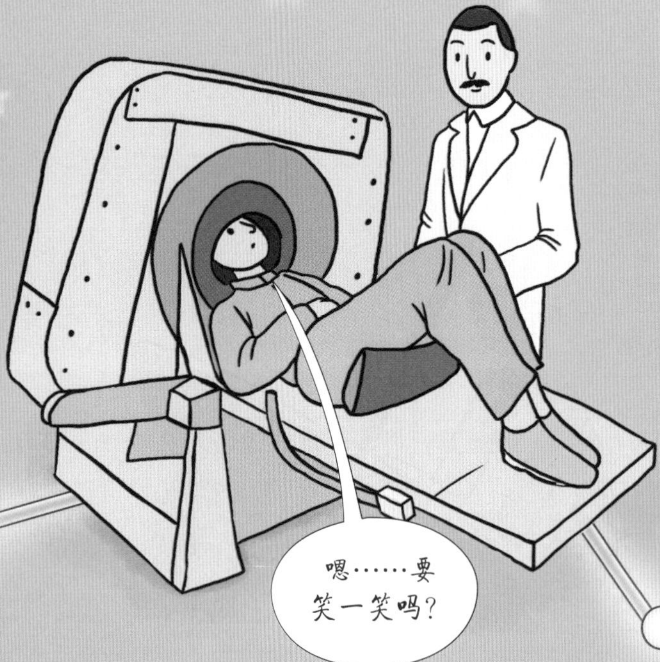

CT扫描仪

计算机断层造影（CT）扫描仪看上去有点像巨大的甜甜圈，可以给人体拍摄分层照片，由英格兰工程师戈弗雷·豪恩斯菲尔德于20世纪70年代早期发明。他先用一个死人的大脑进行试验，再用一个牛脑试验，最后在自己身上试验！

嗯……要笑一笑吗？

磁共振成像

亚美尼亚裔美国人雷蒙德·达马迪安发明了磁共振成像（MRI）技术。1977年，雷蒙德证明，对人体内部的钾施加电波以后，用扫描仪测量钾的活动，就能“看到”癌症肿瘤。每年，放射科医师都要做6000多万次磁共振成像扫描，挽救了无数人的生命。

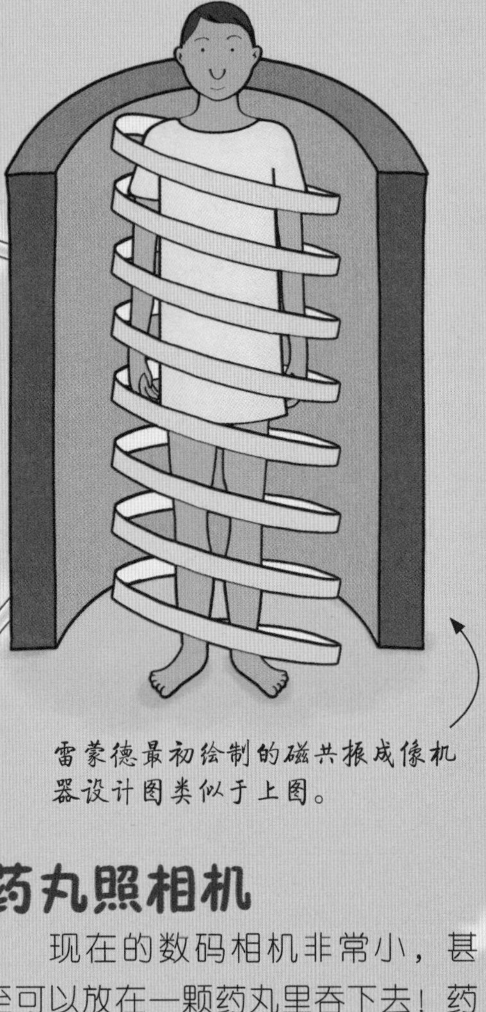

雷蒙德最初绘制的磁共振成像机器设计图类似于上图。

超声

20世纪50年代，伊恩·唐纳德和汤姆·布朗发明了超声，用于检查船舶是否存在缺陷。今天，超声被用来检查胎儿。

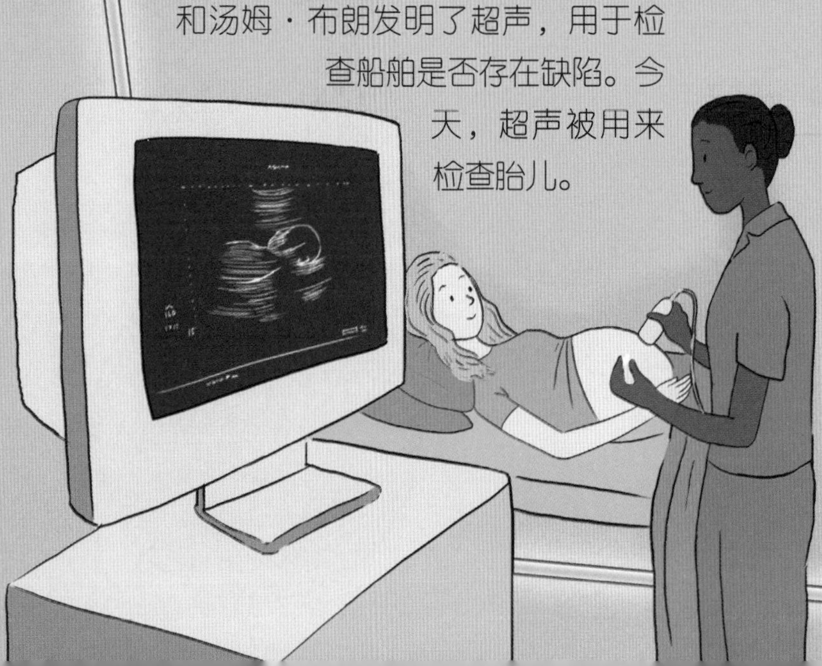

药丸照相机

现在的数码相机非常小，甚至可以放在一颗药丸里吞下去！药丸照相机（也称“胶囊内镜”）通过我们的消化系统时，可以发现内脏的问题。以色列人葛瑞尔·爱登于1997年发明了这种相机，他是在研究导弹（可以定位目标的火箭）时学到的这项技术。

Pillcam

声呐和雷达

声呐是“声音导航与测距”的简称，是一种我们从大自然学来的技术：声波碰上物体反射回来时可以生成图像，很像蝙蝠捕蛾子或者海豚捕鱼的方式。雷达也接收回声，但用的是无线电波，而不是声波。

声呐实验

1822年，丹尼尔·科拉东和查尔斯·斯蒂尔姆利用火药和铃铛在瑞士一个湖里测量声音在水里的传播速度。（声音在水里的传播速度比在空气中快四倍。）这可以说是最早的声呐。

海难

1912年，泰坦尼克号远洋轮船发生沉船事故，导致1,500多名乘客和船员丧生。后来，专家们意识到，如果这艘船安装了声呐，就很有可能看到航道上的冰山，从而避免相撞。

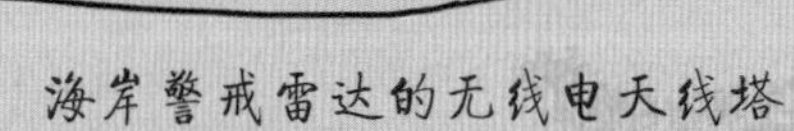
海岸警戒雷达的无线电天线塔

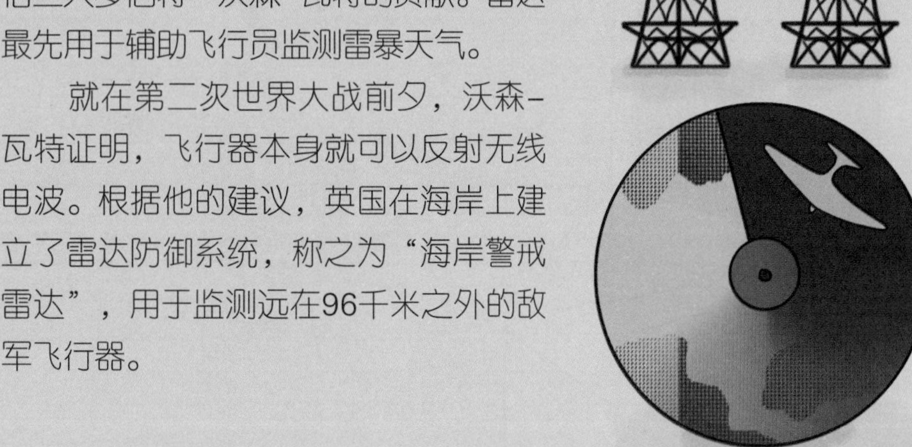

战争时期的雷达

把雷达（“无线电探测和测距”）变成战场上克敌制胜的科技，离不开苏格兰人罗伯特·沃森-瓦特的贡献。雷达最先用于辅助飞行员监测雷暴天气。

就在第二次世界大战前夕，沃森-瓦特证明，飞行器本身就可以反射无线电波。根据他的建议，英国在海岸上建立了雷达防御系统，称之为“海岸警戒雷达”，用于监测远在96千米之外的敌军飞行器。

冰山监测设备

加拿大工程师雷金纳德·费森登设计了一种声呐系统，可以监测到3.2千米外的冰山，可以测量海的深度，还可以用莫尔斯电码发送信号。这项发明后来被应用于潜艇。

测绘海底地图

如今，声呐能帮助船只“看见”海底，即使海底很深而且漆黑一片。人们利用声呐研究岩层以及寻找沉没的船只或飞机。下图这艘船在使用多波束声呐发送扇形信号，信号的回波能描绘出海底的形状。

近炸引信

第二次世界大战期间，近炸引信是同盟国使用的又一种秘密武器。这是一种雷达，安装在发射出去的炮弹弹头内。辐射波监测到目标（比如在空中飞行的飞机）后，就会引爆炮弹。

导航设施

今天，雷达被广泛用于非军事目的，比如在夜间给船只导航。船只的微波雷达扫描仪发送无线电波，然后接收它们遇到固体后反射回来的电波，所得结果会显示在屏幕上。

神奇的激光

如果你已经读到这里，说明你的阅读能力很强。说实在的，光是一种很复杂的东西。现在，我们要讨论真正棘手的主题了：激光！我们每天都会用到激光，比如CD和DVD播放器、激光打印机以及激光显示器，这些设备里都配有激光装置。那么是谁发明了激光，激光的工作原理又是怎样的呢？

认识光子

了解一点关于光的知识，有助于我们理解激光。光是一种电磁辐射，可以说是无形X射线的有形表亲。1905年，德国科学家阿尔贝特·爱因斯坦这样描述道：光是由叫做光子的微小粒子组成的，常以波形曲线（即波包）表示。

激光器的工作原理

激光器中心是一个圆柱体，里面充满一种增益介质：气体或水晶。圆柱体两端各有一面镜子，其中一面镜子上有个小孔。增益介质是由电或光能（光子）“激发”的。光子激发增益介质，让增益介质产生更多光子。最终，光子会从激光器一端射出来。

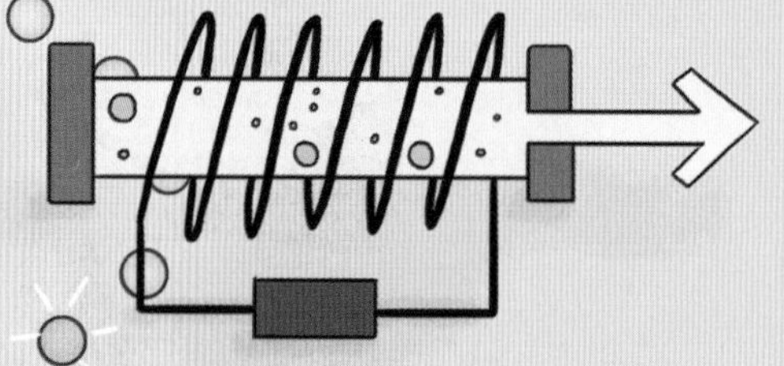

激光器的制造者

1960年，西奥多·梅曼在美国制造了第一个实用的激光器。但如果没有美国、俄罗斯和伊朗等国的其他科学家在20世纪50年代做的开创性工作，梅曼无法造出这个激光器。以下是其中几位为激光器的发明做出贡献的科学家。

超级切割机

激光曾被描述为“一种寻找问题的解决方案”，如今被广泛应用于各类设备中，例如用于切割的激光能切断金属、木材和织物。无论是非常薄的精密零件，还是12毫米厚的钢材，激光都能切割。左图中的切割模式是由电脑控制的。

激光雷达枪

警察用激光雷达枪检查危险驾驶行为。激光雷达的工作原理有点像雷达，但用的是激光，而不是无线电波。激光雷达非常精确，能在瞬间锁定一辆车，并测量这辆车的速度。如果你看到有激光雷达枪对着你家的车，就叫你爸妈减速吧……

激光扫帚

宇航员面对的危险之一是越来越多的太空垃圾，即围绕地球高速飞行的老旧航天器碎片。如果一块金属击中航天器，就可能引发一场大灾难！将来某一天，美国航空航天局可能会使用“激光扫帚”，即安装在地球上的激光武器，来清除太空垃圾，以免它们造成危害。

脑洞大开的发明

煮沸的尿液

荧光棒其实就是一种化学发光棒，最初的发明者是德国化学家亨尼希·布兰德。1669年，亨尼希在一个烧瓶里装了自己的尿液然后煮开，直到尿液着火，变成发光的磷（后来用于制作火柴的一种化学物质）。有意思吧！

给马戴的眼镜

伦敦的多朗德家族因为发明了望远镜和眼镜的镜片而出名，然而在1893年，这家人把他们的发明又往前推进了一步——多朗德先生发明了给马戴的双光眼镜。为什么呢？因为这种眼镜可以使路面看起来比实际更近一些，这样一来，马就会抬高脚步走路。在当时，人们觉得马这样走路很优雅。显然，马非常喜欢戴这样的眼镜。

给鸡戴的眼镜

不光马有眼镜！1903年，美国人安德鲁·杰克逊发明了给农场里的鸡戴的小眼镜。他发明这种眼镜是为了阻止鸡之间相互啄咬，因为这样鸡会生病。这种眼镜的镜片是玫瑰色的，主要作用是保护鸡的眼睛，以免鸡看到血就发狂，啄咬得更厉害。杰克逊的这个发明很管用……

鸽子照相机

还记得那些航空摄影师吗？嗯，1908年，德国的尤利乌斯·诺伊布龙纳博士发明了一种迷你照相机……是给鸽子用的。这种照相机可以绑在鸽子的胸部，通过定时器控制，定时拍摄照片。诺伊布龙纳在公共集市上展示他发明的相机，人们可以现场购买由鸽子拍的照片制作的明信片。后来，诺伊布龙纳的鸽子照相机还在第一次世界大战中派上了用场。听起来不可思议，但这是真事。

躺着也能看书

汉布林眼镜是在20世纪30年代发明的，这种特殊的眼镜能以45度角反射光线。戴上这种眼镜，你平躺着（假设你生病了，或者只是累了）也能看放在你胸口上的书。如果你想买一副，那你运气不错：现在还能在网上买到这样的眼镜呢！

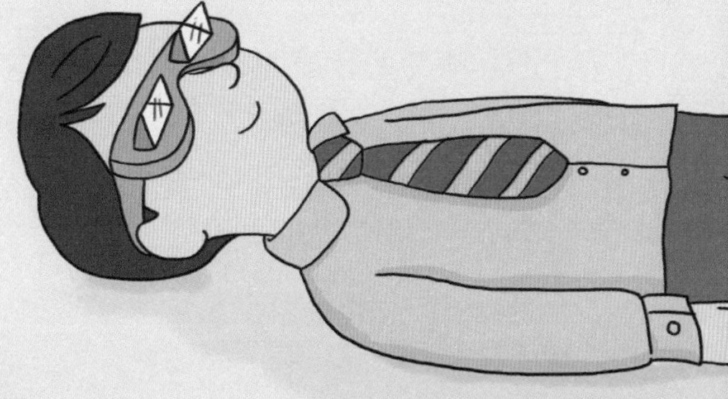

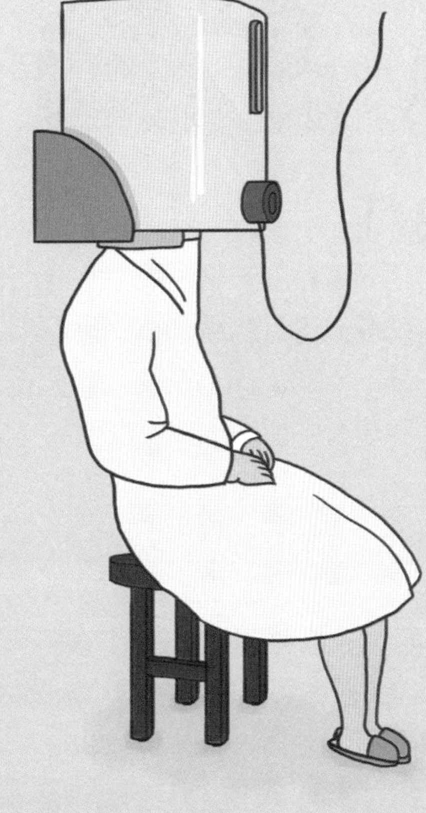

光照治疗法

从19世纪开始，日光浴就成了头部疾病——如流鼻涕和耳朵疼——的流行“治疗方法”。左图这种治疗模式从20世纪30年代开始流行：病人把头放到一个大罐子里，里面的紫外线射线枪会把人造太阳光射在病人头上。紫外线可以帮助人体产生有益于人体健康的维生素D。但是紫外线也能烧伤皮肤，引发癌症，所以也是有危害的。

反飞行器大号

这种东西看起来像一支巨大的铜管乐队，但实际上是巨大的喇叭状听筒。两次世界大战中军队都使用了这种听筒，来监听靠近的敌军飞行器发动机的声音。罗伯特·沃森-瓦特研发出雷达后，这种军用大号就没有用武之地了。

高明和不高明的预测

电灯泡

19世纪70到80年代期间，托马斯·爱迪生在研究电灯泡时，很多人认为他根本不可能成功。以下是这些人说的一些蠢话：

“凡是熟悉这一学科的人都会把这视为显而易见的失败。”
——亨利·莫顿，美国史蒂文斯理工学院校长

“巴黎博览会闭展时，电灯也会随之熄灭，从此以后无人知晓。”
——牛津大学教授伊拉斯谟·威尔逊

“电灯没有未来。”
——英国发明家约翰·佩珀

打开电灯

虽然佩珀对电灯的未来进行了悲观（而且明显错得离谱）的预测，但他仍热衷于研究电。1863年，在威尔士王子爱德华与丹麦公主亚历山德拉的婚礼上，他用弧光灯点亮了伦敦特拉法尔加广场和圣保罗大教堂。他还发明了一种将演员的“灵魂”投射到舞台上的方法，手段非常高明，甚至连杰出的科学家迈克尔·法拉第也央求他解释其中的奥秘。

但佩珀也有不那么聪明的时候。1882年的夏季十分干旱，当时他在澳大利亚，想用大型舰炮和火箭弹向天空开炮来造雨。但他失败了，受到众人嘲笑，于是只好作罢。

是的，这些预言真的实现了……

有些人比较擅长预测。法国作家朱尔·凡尔纳在他1863年写的小说《二十世纪的巴黎》中，针对1960年提出了几个精准的预测：

★ 有电灯的城市
★ 摩天大楼
★ 高速火车
★ 全球信息网络（很像因特网）
★ 传真机
★ 汽车
★ 死刑电椅（可怕！）
★ 大规模杀伤性武器（不好！）

让人吃惊的是，凡尔纳的出版商埃策尔先生认为他写的故事太无聊了，因此直到凡尔纳去世，这本书都没有出版。

西蒙说

占星家通过研究星图和黄道十二宫图来预测未来，但有时候会出错。天文学家跟他们相比很不同。天文学家是太空科学家，但有时候也免不了犯错。

在古希腊，亚里士多德认为太阳是围绕地球运转的。后来，托勒密支持了他的说法。现在我们当然知道事实恰好相反：在太阳系中，行星是围绕太阳运转的。

再后来，到1898年时，加拿大裔美国天文学家西蒙·纽科姆说：“我们很可能在接近天文学知识的极限。”经过几年研究，西蒙非常明智地改变了自己的想法，将天文学称为“一个不可穷尽的领域”。

索引

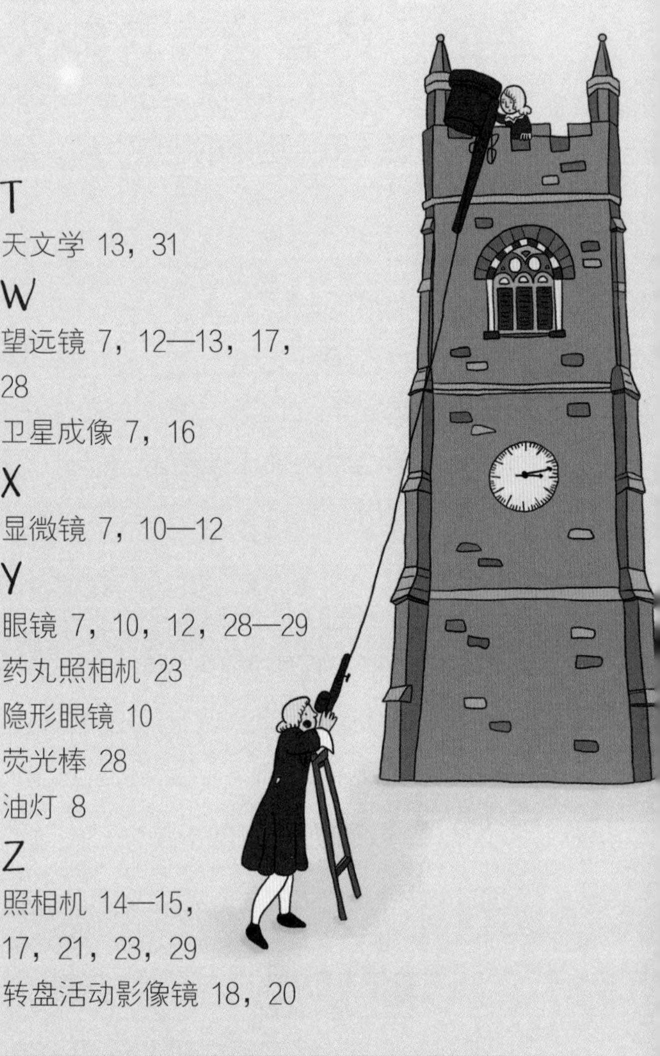

作者简介

马特·特纳出生于英国，20世纪80年代毕业于拉夫伯勒大学艺术学院，毕业后一直担任图片研究员、编辑和作者。他的书题材广泛，涉及博物学、地球科学和铁路等，并为百科全书和分册出版的丛书写过数百篇文章，从大象到抽象艺术无所不包。他现在和家人住在新西兰奥克兰附近，他还是当地海岸警卫队的志愿者，平时也涉猎工艺品的制作。

绘者简介

萨拉·康纳生活在英格兰美丽的乡村，栖身于一座可爱的小木屋里，有几只狗和一只猫做伴。她平时用画笔描绘身边的世界。她总能从大自然中获得灵感，这对她的很多作品都有影响。萨拉之前用钢笔和颜料画插图，但近些年，她开始在电脑上绘画，因为这更适应今天的产业发展。然而，她还是喜欢时不时拿出水彩颜料来，画一画自己花园里的鲜花！

INCREDIBLE INVENTIONS

ALL ABOUT LIGHT

Written by Matt Turner
Illustrated by Sarah Conner

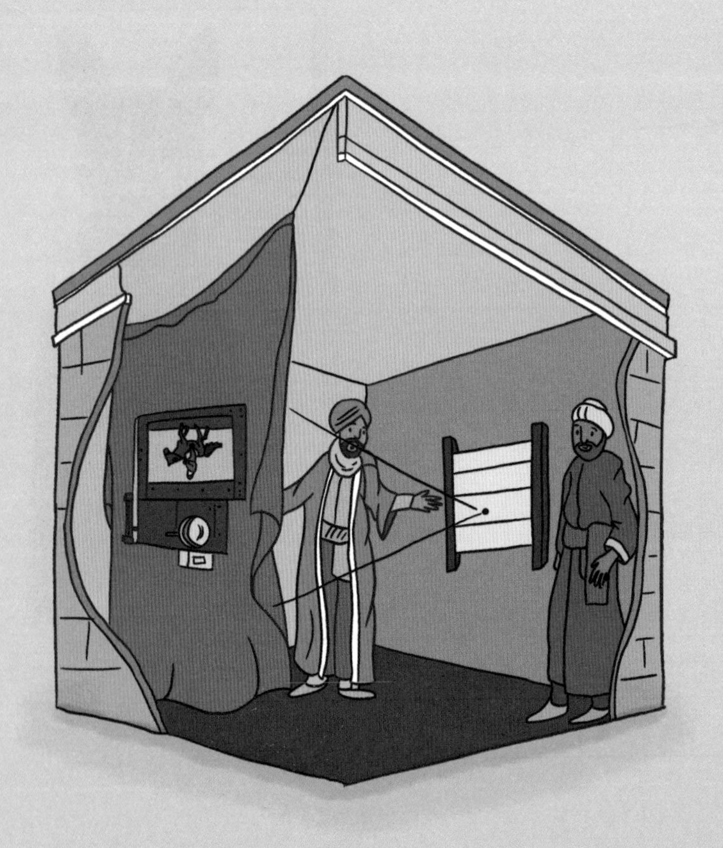

外语教学与研究出版社
FOREIGN LANGUAGE TEACHING AND RESEARCH PRESS
北京 BEIJING

Before magnifying glasses, how did people look at things big size? Through bowls of water, of course!

Then microscopes and telescopes came along... Follow the trail of inventions and devices that have enabled us to see everything from stars to cells, and what's totally hidden from sight. Discover how

- Thomas Edison tested 3,000 light bulb designs before finding 'the one'
- the camera began life as a darkened room (with a hole in the wall)
- movie-goers found Cinerama so exciting, they screamed
- MRI scanners 'see' cancer tumours
- lasers can cut through steel

and so much more. There are also crazy inventions (spectacles for horses, why not?) and off-beam predictions – some people thought electric light would never catch on...

Galileo (1564–1642) built telescopes to look at objects both close-up and far-off, from insects to planets.

CONTENTS

INTO THE LIGHT

When our early ancestors first learned to make fire and control it, they altered human destiny. Now able to make light (and heat), they no longer had to go to bed at sundown. They could stay warm at night, and cook food. No more raw meat!

3D projector, 1890s

Eating better food may even have been a 'trigger' that led to our larger brain size, making us cleverer. Fire also helped the early humans keep pesky bugs and wild beasts at bay. And it helped them hollow out canoes for travelling the oceans and finding new lands. Of course, these advances happened a long

Magnifying glass bowl, at least 2,000 years ago

time ago, and we can't say who invented them. They stand more as 'big leaps forward' in human society.

As well as making light, we've learned to focus it through a lens. When humans first polished lenses from rock crystals, or used glass globes full of water, they found that these simple lenses could 'bend' light rays and show things in close-up. Light is the important bit in microscopes, telescopes, spectacles and movie cameras.

Satellite imaging, 1959

Come with us on a roundabout journey from those early fire-making days to the modern era of lasers and satellite photography. We'll explore some of the more unexpected ways of looking at things – such as radio waves, microwaves, X-rays and sound waves, which have given us radar and medical scanners.

And we'll poke 'light' fun at some of the crazier inventions people have dreamed up – such as spectacles for chickens and horses!

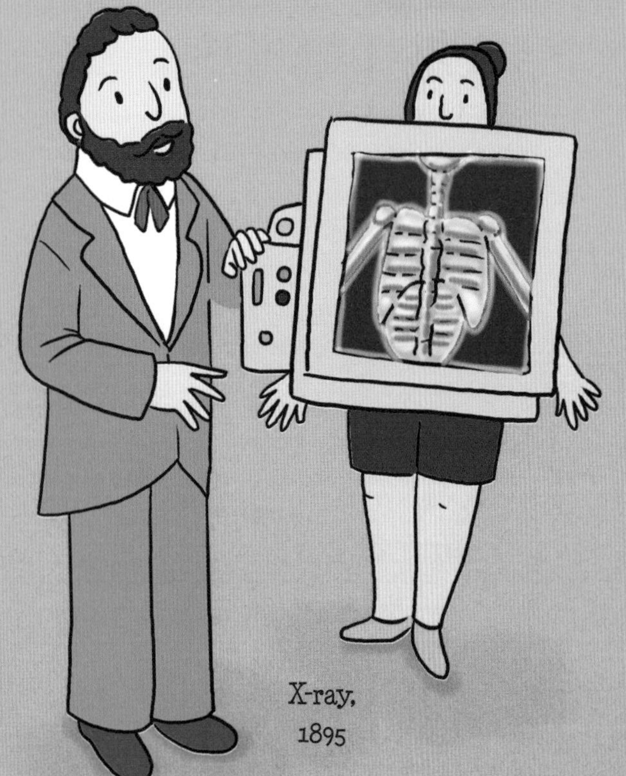

X-ray, 1895

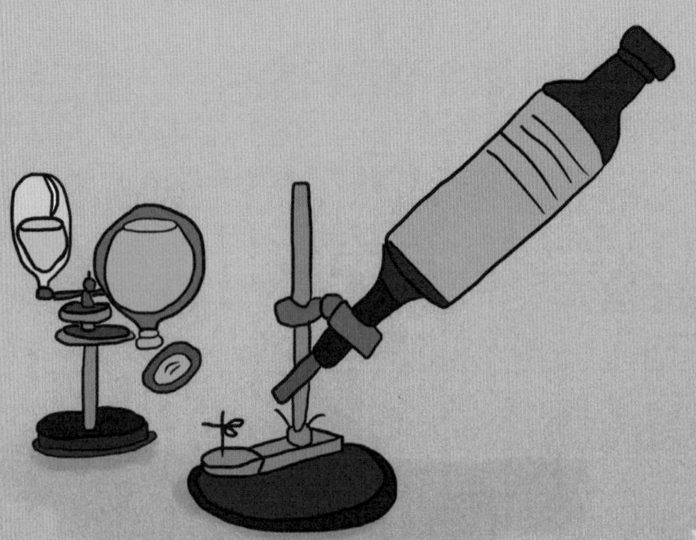

Hooke's microscope, 1665

MAKING LIGHT

You probably have an electric light on right now, don't you? Flicked it on with a switch? Our ancestors weren't so lucky: they used all sorts of tricks, from rubbing sticks together to making matches or giant batteries, all in order to create light.

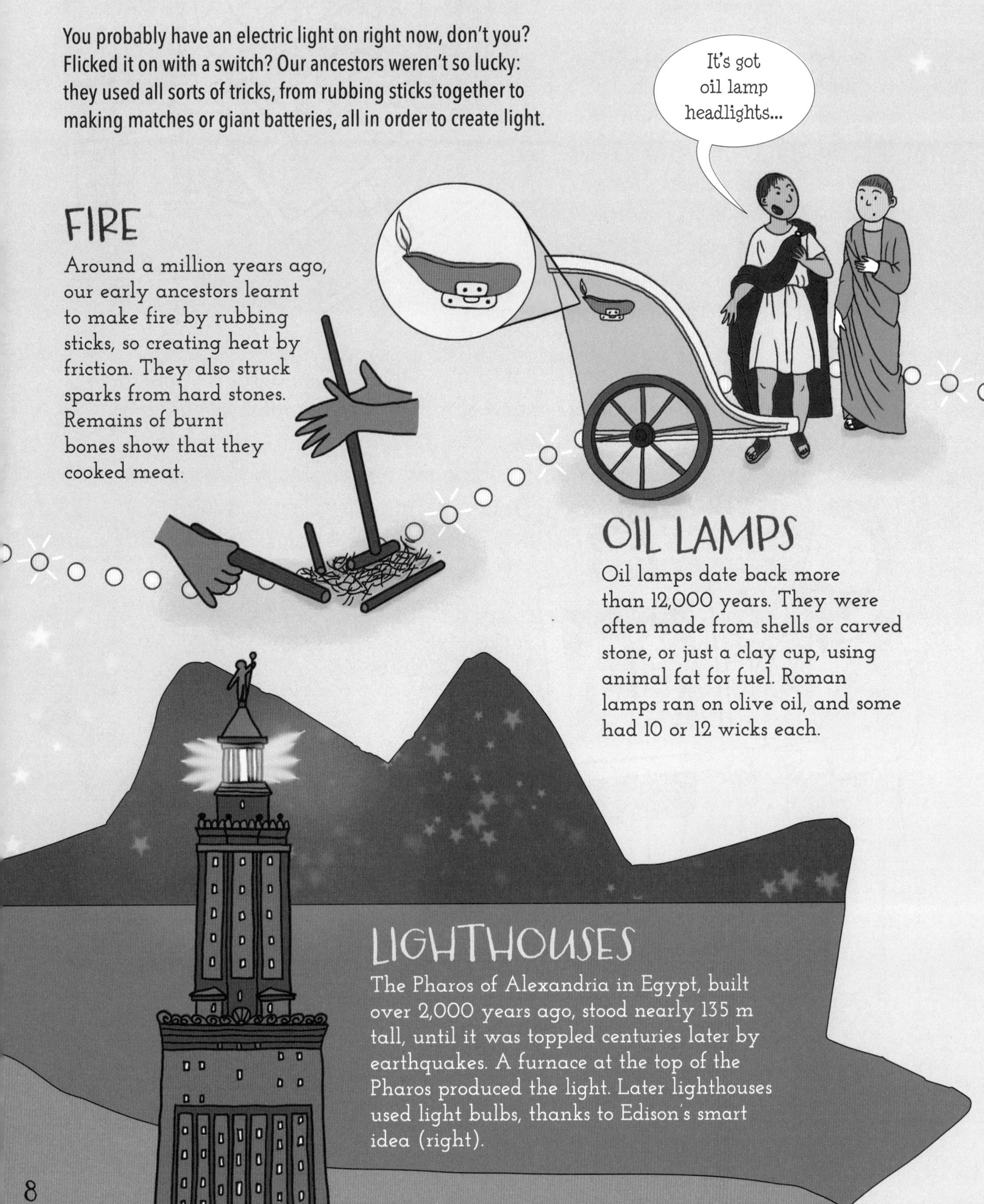
It's got oil lamp headlights...

FIRE

Around a million years ago, our early ancestors learnt to make fire by rubbing sticks, so creating heat by friction. They also struck sparks from hard stones. Remains of burnt bones show that they cooked meat.

OIL LAMPS

Oil lamps date back more than 12,000 years. They were often made from shells or carved stone, or just a clay cup, using animal fat for fuel. Roman lamps ran on olive oil, and some had 10 or 12 wicks each.

LIGHTHOUSES

The Pharos of Alexandria in Egypt, built over 2,000 years ago, stood nearly 135 m tall, until it was toppled centuries later by earthquakes. A furnace at the top of the Pharos produced the light. Later lighthouses used light bulbs, thanks to Edison's smart idea (right).

MATCHES

The Chinese invented matches about 1,000 years ago, naming them 'fire inch-sticks'. Reliable friction matches – lit by scraping on sandpaper – first appeared in Britain in the 1820s. These early matches contained phosphorus, which made match factory workers – and sellers, like this boy – very sick.

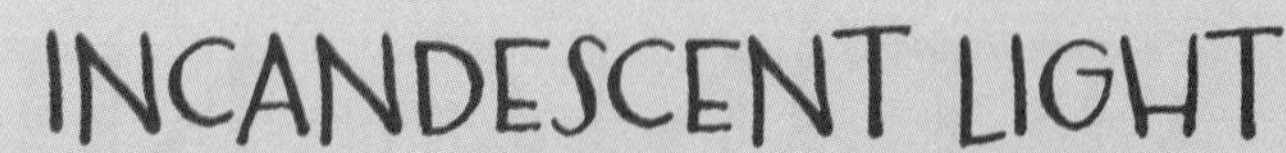

INCANDESCENT LIGHT

British scientist Humphry Davy invented incandescent light (that is, light resulting from heat) in 1801. He heated a strip of platinum till it glowed, using the world's (then) most powerful battery: 2,000 linked cells. Wow!

I say, do you have this battery in a pocket torch size?

DAVY'S SAFETY LAMP

Davy is famous for inventing a 'safety lamp' in 1815, for use by coal miners. Its flame was covered over, to reduce the risk of gas explosions. It didn't work too well: the light was dim, and the explosions continued. Humph!

Hey, I have a great idea...

EDISON'S GREAT IDEA

American Thomas Edison didn't invent the first light bulb, but in 1879 he came up with the first reliable one – after testing more than 3,000 bulb designs and about 6,000 different filaments! (The filament is the glowing thread inside.)

SEEING NEAR

Magnifying glasses, spectacles, contact lenses and microscopes all rely on the fact that a curved piece of glass – a lens – focuses light and makes things look closer than they are. Through history, lenses have led to many scientific discoveries (as well as giving us one more thing to lose on the bus).

Carrots, frogs' legs, milk...

MAGNIFIER

Before magnifying glasses existed, the ancient Romans just filled a glass bowl with water. Looking through it, they saw things at larger size.

SPECTACLES

Early Arabs knew about optics, but in the West, English monk Roger Bacon was the first to write about lenses, in 1268. Within 20 years, the Italians had invented 'clip-on' specs.

I wish I couldn't see these frogs' legs.

BIFOCALS

American statesman Benjamin Franklin (1706-90) invented bifocals: spectacles with lenses of two different strengths. On a trip to France, he used them to see both his dinner and fellow-diners.

CONTACT LENSES

In 1888, German doctor Adolf Fick designed the earliest practical contact lenses. He first tested them on rabbits. (But how did he tell that they worked?!)

How many carrots am I holding up?

ELECTRON MICROSCOPE

Super-powerful scanning electron microscopes (SEMs) were invented in the 1930s by German TV engineer Max Knoll. SEMs were first used to look closely at metals. They also take terrifying pictures of bugs...

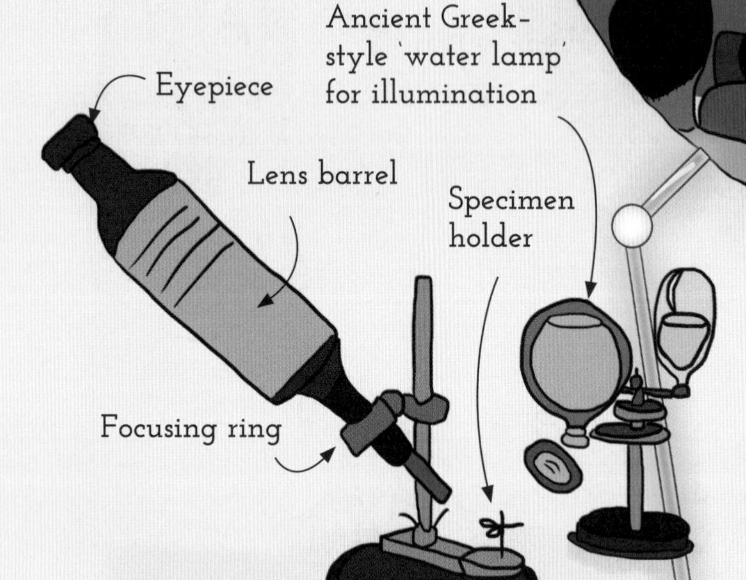

HOOKE'S MICROSCOPE

Englishman Robert Hooke was the first to use the word 'cell' to describe the basic unit of life. He looked at cells through his microscope of 1665.

SEEING THE INVISIBLE

A microscope uses really powerful lenses to make truly tiny things look big. The first were compound microscopes: tubes with sets of lenses that worked together to magnify. With these, scientists saw bacteria, blood cells and yeast for the first time.

This scope (top view below, side view left) was designed by Dutchman Antonie van Leeuwenhoek (1632-1723). You put a specimen on a sharp point and viewed it through the 275x lens mounted in the plate.

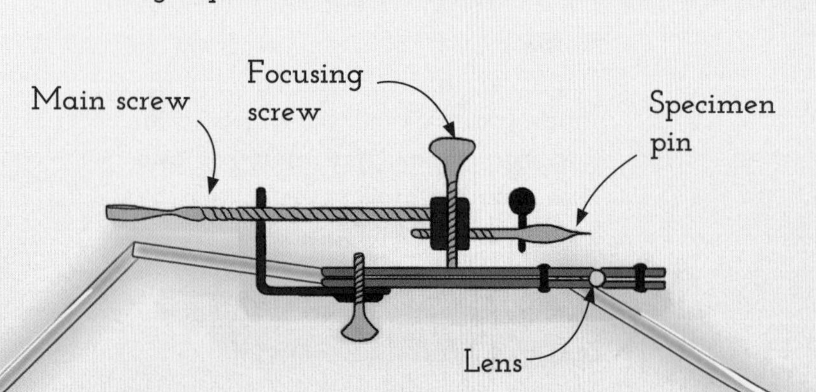

SEEING FAR

Telescopes have enabled astronomers to look deep into space and make discoveries about planets, moons and stars. Early telescopes used lenses, rather like microscopes, but today we also use radio telescopes to explore the universe.

EARLY TELESCOPE

Dutchman Hans Lippershey made magnifiers and ran a spectacles shop. In 1608, he came up with an early telescope, which he called a 'looker'. He offered it to the government to use in battle; they promised him a reward if he converted them into binoculars!

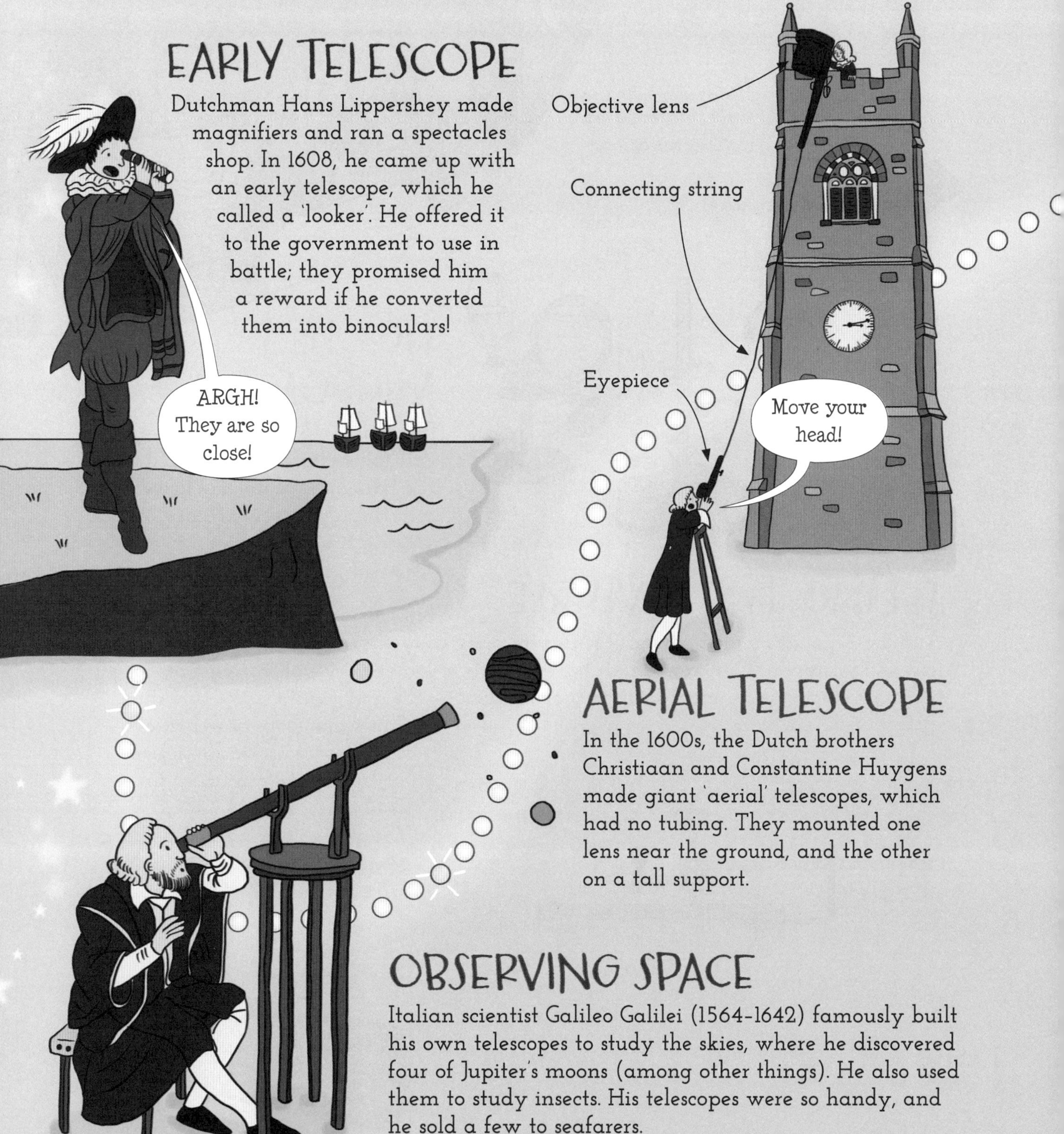

ARGH! They are so close!
Objective lens
Connecting string
Eyepiece
Move your head!

AERIAL TELESCOPE

In the 1600s, the Dutch brothers Christiaan and Constantine Huygens made giant 'aerial' telescopes, which had no tubing. They mounted one lens near the ground, and the other on a tall support.

OBSERVING SPACE

Italian scientist Galileo Galilei (1564-1642) famously built his own telescopes to study the skies, where he discovered four of Jupiter's moons (among other things). He also used them to study insects. His telescopes were so handy, and he sold a few to seafarers.

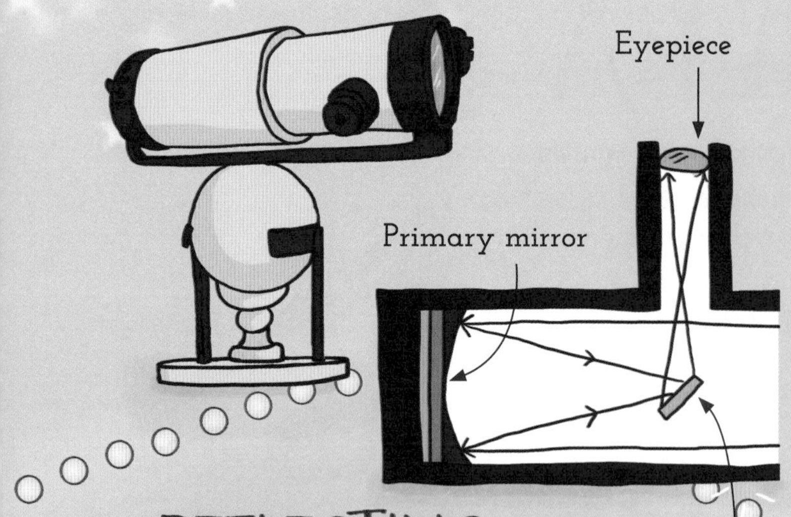

FINDING SATURN'S MOONS

In 1789, German-British astronomer William Herschel built this 12-m-long giant. It was a reflecting 'scope, like Newton's, but with an improved mirror set-up. He used it to look at Saturn's moons. One of them, Mimas, looks scarily like the Death Star...

REFLECTING TELESCOPE

Early telescope lenses tended to split light into rainbow colours, making images hard to see. British scientist Isaac Newton fixed this problem in 1668 with a reflecting telescope. This used a mirror, not a lens, to capture the image.

RADIO TELESCOPE

Modern radio telescopes don't 'look' with lenses, but 'listen' like TV antennae. American Karl Jansky built the first radio telescope in a potato field in 1931. He attached antennae to a radio receiver to detect radio waves from space – energy from star activity – which he heard as a steady hiss.

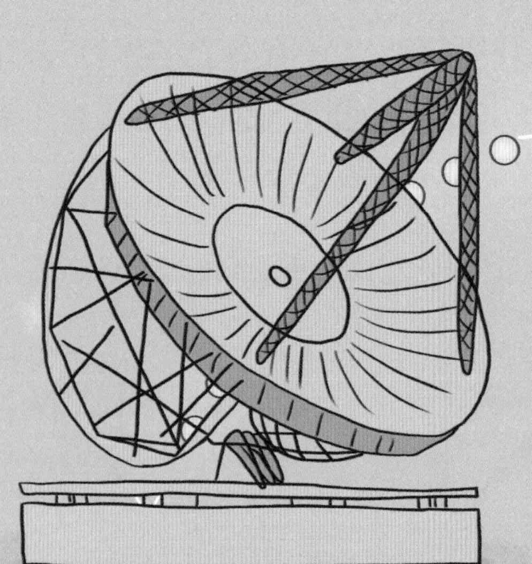

RADIO ASTRONOMY

Jansky's discovery inspired Grote Reber, a fellow American, to build a radio telescope in his back garden in 1937. Reber used it to create a 'map' of space radio signals. In those early years, he was the world's only radio astronomer! Today there are thousands...

THE CAMERA

Cameras are amazingly complex today, but the basic idea comes from the 'camera obscura' (meaning 'darkened room'), or pinhole camera. This is a box, with the only light coming from a small hole in one wall. Light enters and, like a film projector, casts an image (upside down) on the opposite wall. If you line that wall with photosensitive film or paper, you capture the image as a photograph.

CAMERA OBSCURA

The Arab scientist Alhazen (965–1039) described how pinhole cameras worked. Astronomers and artists took up the idea.

PHOTOGRAPHIC PAPER PICTURES

Around 1800, British scientist Thomas Wedgwood experimented with photosensitive paper. He left it on sunny windowsills, taking 'shadow photos' of leaves and other objects.

EARLY PHOTOGRAPHY

The true inventors of photography were Frenchmen Nicéphore Niépce and Louis Daguerre. Nic tried photosensitive chemicals (silver chloride, bitumen, lavender oil) and began taking basic photos around 1816. In 1839, Daguerre developed the 'daguerreotype' method. This used a polished silver sheet coated with chemicals and exposed to light in a camera obscura called, yep, a 'Daguerre' (below). It took very sharp photographs.

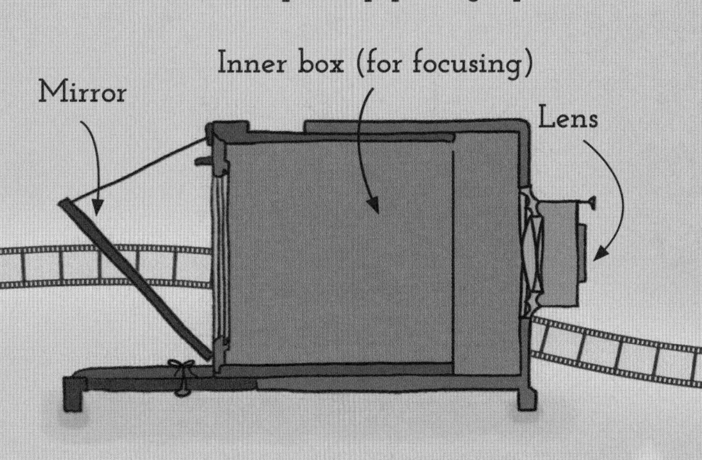

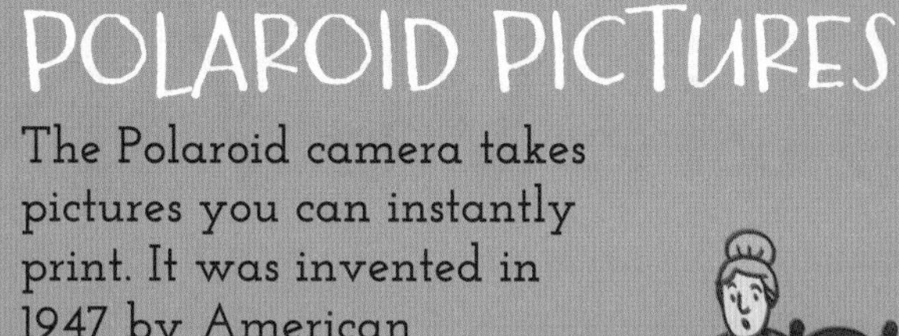

POLAROID PICTURES

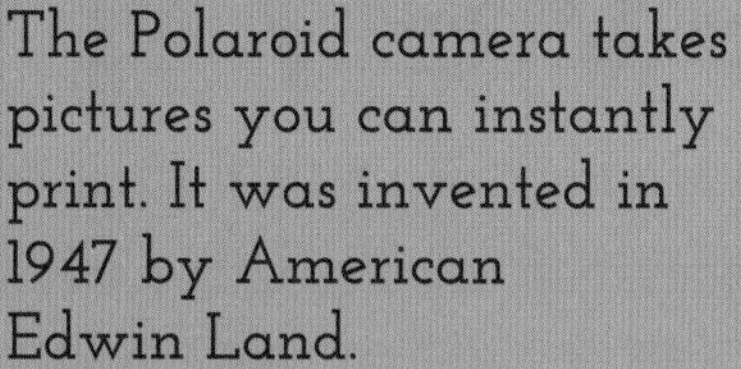

The Polaroid camera takes pictures you can instantly print. It was invented in 1947 by American Edwin Land.

DIGITAL CAMERAS

The invention of the charge-coupled device (CCD), an electronic gizmo, by Willard Smith and Bill Boyle in 1969 led to the end of film and the dawn of the digital camera, invented in 1975 by Stephen Sasson.

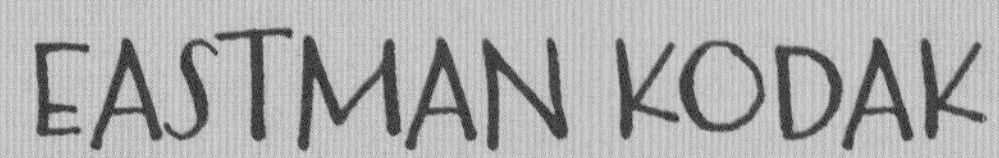

EASTMAN KODAK

The big revolution in photography came with the invention of film by American George Eastman in 1884. It featured in his Kodak camera, which he patented in 1888. The invention of small, affordable cameras, such as the Brownie, meant that, from now on, anyone could be a photographer.

POSITIVE AND NEGATIVE

Also in the 1830s, British scientist William Fox Talbot came up with the calotype: a better, faster method than the daguerrotype. With the calotype, you could make a negative – a 'back to front' image – and use this to make positive copies.

SURVEYING

Surveying means measuring the land. It is important because it gives us maps and GPS (which stops mum and dad getting lost when driving). It also led surveyors on some exciting and dangerous journeys.

MEASURING MOUNTAINS

Ancient Egyptians used long, knotted ropes for measuring distances. Early Chinese (below) used a protractor on a pole to calculate mountain heights by measuring angles. This method, or versions of it, saw use for many centuries.

I'm sure it's around here somewhere...

FIRST MAP

Here is the first known map of the world: the Imago Mundi from Babylon (in modern-day Iraq). An unknown person scratched it onto a clay tablet some 2,500 years ago. The ring represents the ocean, surrounding the lands.

SATELLITE IMAGING

Nowadays, satellites orbiting Earth take photos so clear, they can spot you sunbathing in the garden. Explorer 6, launched in 1959, took some of the first photos from space. They were very fuzzy – but it was the dawn of satellite imaging.

THEODOLITES

To chart new empires, surveyors went out with theodolites. These were basically telescopes that measured angles very accurately. Some were huge, like this one built by Jesse Ramsden in England in the 1780s, weighing nearly 100 kg. And in the young US, the young George Washington was the first surveyor of northern Virginia...

AERIAL PHOTOGRAPHY

Frenchman Nadar launched aerial photography in 1858. This was his biggest balloon, named The Giant. He also invented airmail! During 1870-71, he used a fleet of balloons to smuggle more than 2,000,000 messages out of Paris, which was under siege by the Germans.

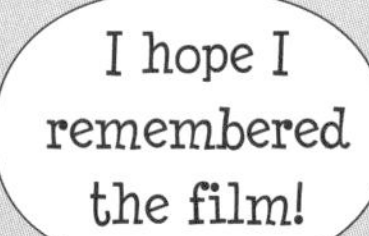
I hope I remembered the film!

ROCKET CAMERAS

Swedish genius Alfred Nobel (famous for the Nobel Prize) pioneered aerial photography from rockets in 1897. In 1906, German engineer Albert Maul (right) built a more efficient design, powered by compressed air.

RECONNAISSANCE

Maul's 'rocket-cam' never really took off, because very soon aerial photographers were using planes. First off was Wilbur Wright (pioneer of flight), in 1903. In World War I, enemy positions were recorded in this way by both sides.

MAKING MOVIES

Who doesn't love watching movies? Your great-great-grandparents, that's who. After all, the film camera is still quite young, having first appeared in the 1890s. But long before then, inventors had begun to make clever toys that tricked our eyes into thinking we were looking at moving pictures… movies.

PHENAKISTOSCOPE

One toy was the phenakistoscope of 1832, invented by Belgian Joseph Plateau. It had two discs, one ringed with pictures. You spun the discs, looked through a slot at a mirror, and saw the pictures 'join up' and move.

ZOOPRAXISCOPE

Similar toys included the zoopraxiscope (Eadweard Muybridge, 1879). It made the first moving photographic image.

FIRST MOVIE CAMERA

Some claim that Thomas Edison invented the movie camera – but the true pioneer was France's Louis Le Prince. He used this camera (right) to shoot some film in Leeds, England, in 1888. Two years later he disappeared, so he never got rich from his invention.

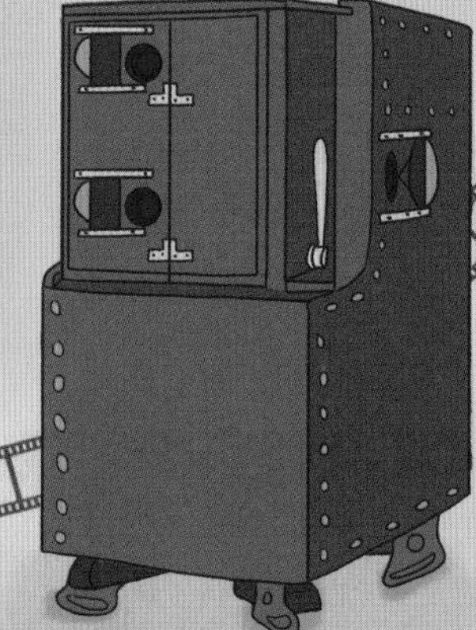

KINETOGRAPH

Edison and his helper Bill Dickson did invent a movie camera called the Kinetograph. And in 1891 he unveiled the Kinetoscope projector, which showed movies. But only one person at a time could use it.

That'll never fit in our living room!
Gosh!
Wow!

VCR TECHNOLOGY

Nowadays you can watch movies on a smartphone. A generation ago, your mum and dad watched them with a VCR, which is short for 'video cassette recorder'. The first VCR was made by Russian-American Alex Poniatoff and his Ampex company. It was so huge you couldn't even carry it, let alone tuck it in your pocket.

MOVIEMAKERS

The first true moviemakers were French brothers Auguste and Louis Lumière. In 1895, they filmed their workers leaving the factory, then gave a public screening of – guess what they called it? – *Workers Leaving the Lumière Factory*. It sounds deadly boring, but viewers were amazed (even though these first films had no sound at all).

VIEWING IN 3D

3D movies are great fun. But they, like the old phenakistoscope, are simply clever tricks that fool your brain. And you might think that making and viewing three-dimensional pictures is new, but it's more than 150 years old – older, even, than the movie camera.

STEREOSCOPE

3D pictures date back almost to the invention of photography. The Brewster Stereoscope of 1849 held two photos – left and right – each taken from a slightly different spot. Viewed together, they gave a 3D effect.

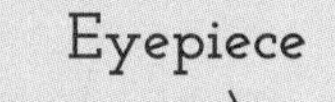

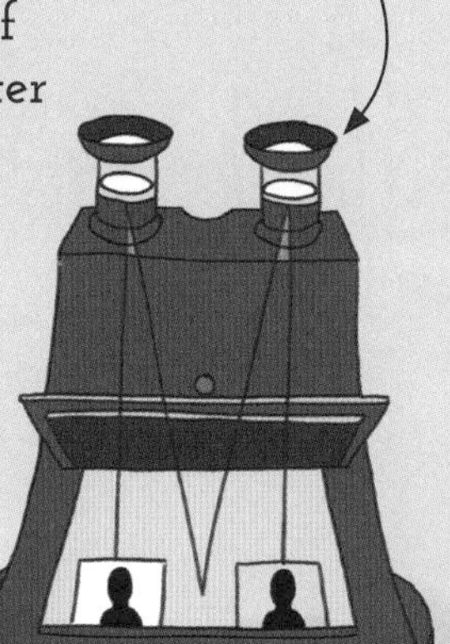

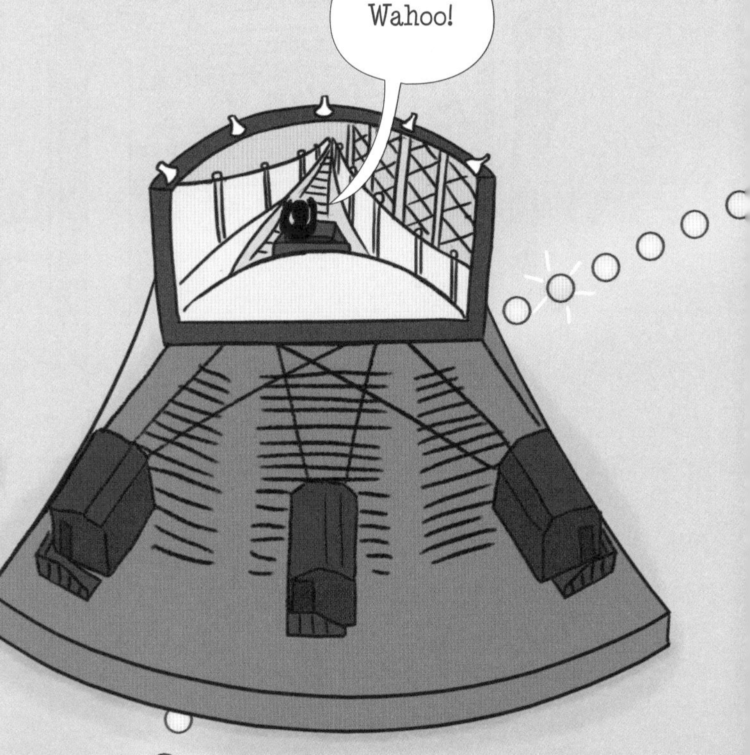

3D PROJECTOR

Alfred Molteni, an Italian-French optician, used a special slide projector with two lenses to create a 3D effect. This audience is watching his nature slides.

CINERAMA

A more modern, movie version of Molteni's rig was Cinerama, invented by American Fred Waller. It used a huge, curved cinema screen, with a 'joined-up' picture created by three film projectors. At the first viewing in 1952, people were so excited, they screamed! But Cinerama was hopelessly complicated and expensive until they worked out a way of using just one projector.

3D SCANNING

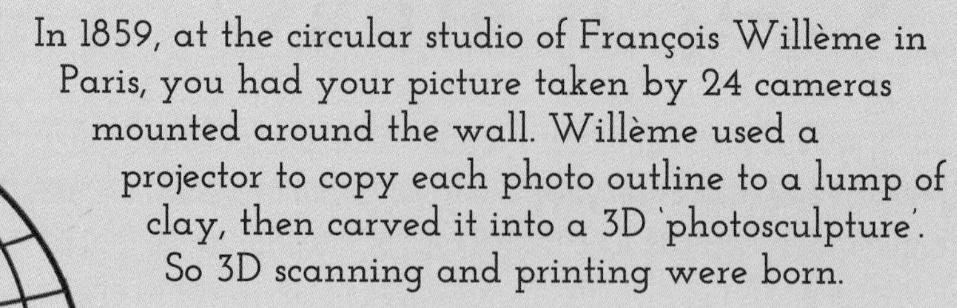

In 1859, at the circular studio of François Willème in Paris, you had your picture taken by 24 cameras mounted around the wall. Willème used a projector to copy each photo outline to a lump of clay, then carved it into a 3D 'photosculpture'. So 3D scanning and printing were born.

I love being the centre of attention!

SPECIAL EFFECTS

The modern version of Willème's rig, as used in the movie industry, is called 'photogrammetry'. Digital cameras take lots of scans of an actor, then animators use these to create special effects, which you watch at the movies.

APP SCAN

123D Catch, created in 2009 by Autodesk, takes 3D scans of objects, which can then be tinkered with on a computer, and even 3D-printed. Willème would have loved it!

SCANNING THE BODY

So far we've looked at bacteria, outer space, planet Earth… now it's time to look inside the body. Sounds gross? Maybe – but it saves millions of lives. Today it's called radiography, or radiology, because it uses the types of radiation wave that can pass through our bodies. Radiographers can also use sound waves to take pictures of our insides.

EARLY ENDOSCOPE

Around 1800, German doctor Philipp Bozzini invented his 'light conductor'. It had a candle and lens for looking right inside an ear or throat. It was one of the first endoscopes.

Oh my !

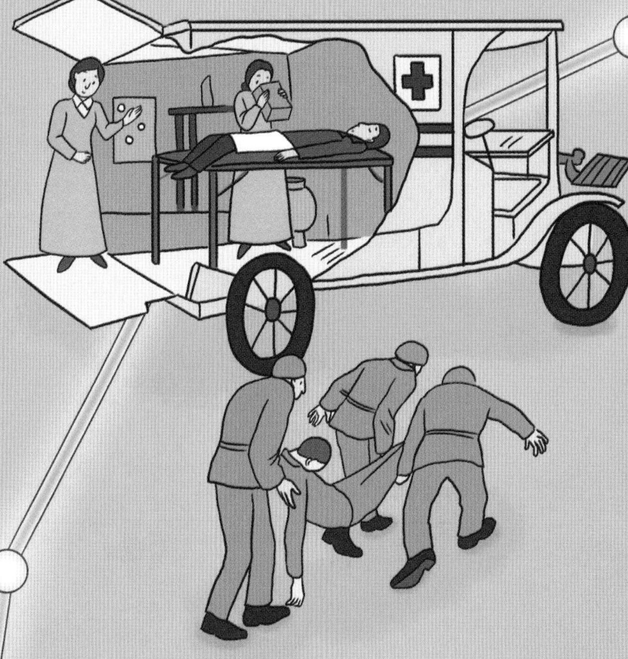

X-RAYS IN BATTLE

Scientists soon realized X-rays could be lifesavers. Polish-French chemist Marie Curie was a brilliant radiologist. During World War I, she sent special medical trucks to the battlefields to take X-rays of wounded soldiers. She won many awards for her studies on radioactivity.

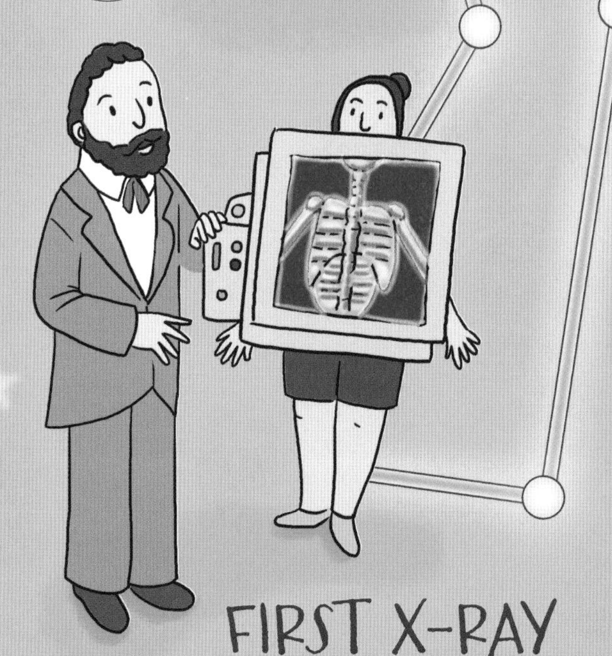

FIRST X-RAY

In Germany in 1895, Wilhelm Röntgen discovered that certain rays could pass straight through cardboard onto a screen, where they glowed. He took the first X-ray photograph, showing the bones in his wife Anna's hand. People realised they could now look inside the human body.

CT SCANNER

The computerized tomography (CT) scanner, which looks a bit like a giant doughnut, photographs the body slice by slice. English engineer Godfrey Hounsfield invented it in the early 1970s. He tested it first on a dead human brain, then a cow's brain – then on himself!

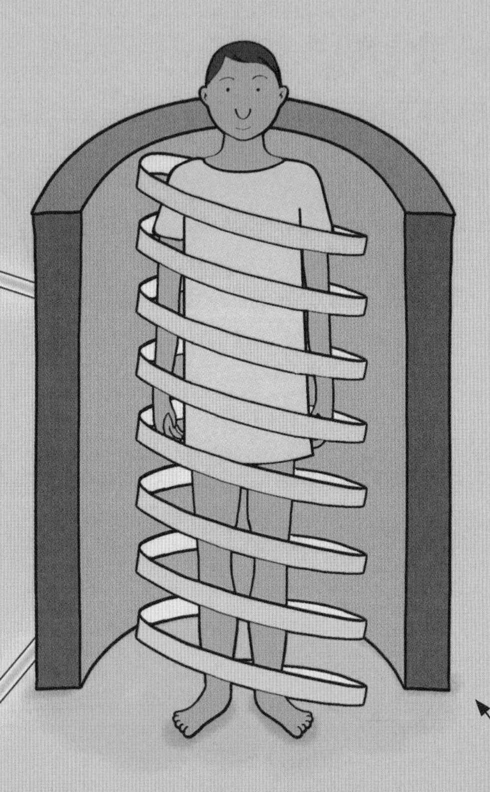

MRI

Armenian-American Raymond Damadian created MRI, or 'magnetic resonance imaging'. In 1977, Ray showed that, by measuring what happens to the potassium inside us when it is energized, a scanner can 'see' cancer tumours. Each year, radiologists take over 60 million MRI scans, helping to save countless lives.

Ray's original sketch for an MRI machine looked a bit like this.

ULTRASOUND

Ian Donald and Tom Brown invented ultrasound in the 1950s for checking ships for flaws. Today it's used for checking babies before they're born.

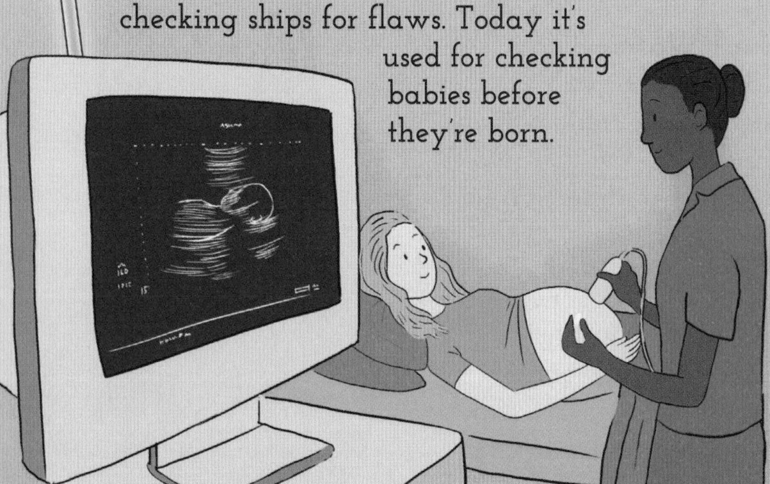

PILLCAM

Digital cameras are now so tiny, you can even swallow one in a pill! The PillCam can spot problems in our gut as it passes through the digestive system. It was invented in 1997 by Israeli Gavriel Iddan. He had learnt his skills by working on guided missiles – rockets that find their own way to a target.

Pillcam

SONAR AND RADAR

Sonar is short for '**so**und **n**avigation **a**nd **r**anging'. It is a technology we've borrowed from nature: it creates pictures from sound waves as they echo off objects – rather as a bat catches moths, or a dolphin catches fish. Radar, too, picks up echoes, but it uses radio waves instead of sound.

SONAR EXPERIMENT

In 1822, Daniel Colladon and Charles Sturm used gunpowder and bells in a Swiss lake to measure how fast sound travelled in water. (It's four times faster than in air.) This was arguably the first sonar.

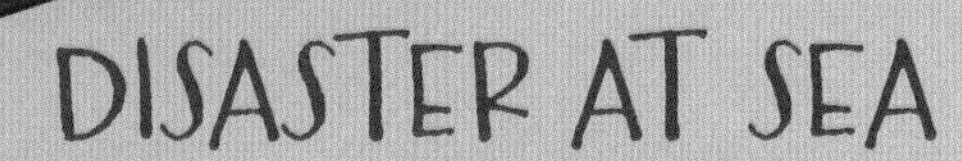

DISASTER AT SEA

In 1912, more than 1,500 passengers and crew died when the *Titanic* ocean liner sank. Experts later realized that, if the ship had been fitted with sonar, it would likely have seen the iceberg in its path, and avoided it.

WARTIME RADAR

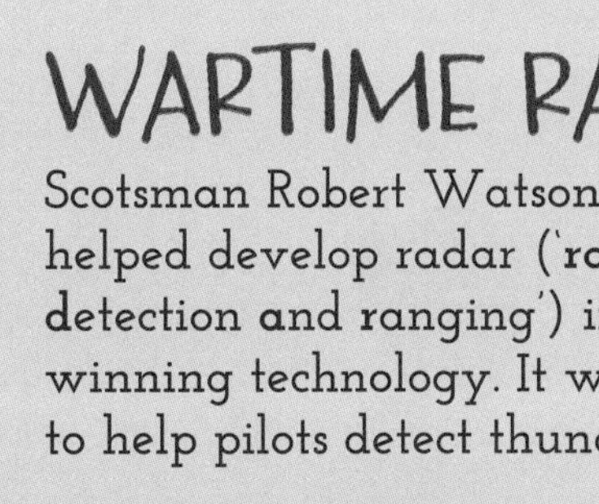

Scotsman Robert Watson-Watt helped develop radar ('**ra**dio **d**etection **a**nd **r**anging') into a war-winning technology. It was first used to help pilots detect thunderstorms.

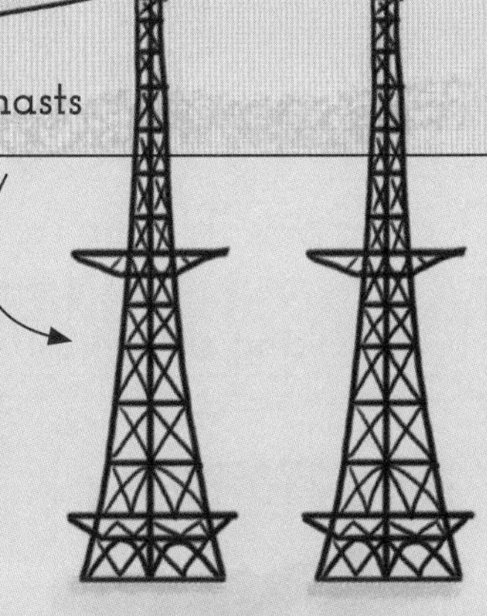
Chain Home radio masts

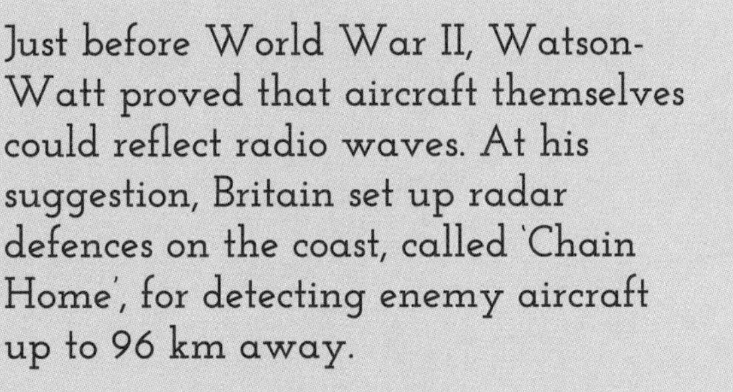

Just before World War II, Watson-Watt proved that aircraft themselves could reflect radio waves. At his suggestion, Britain set up radar defences on the coast, called 'Chain Home', for detecting enemy aircraft up to 96 km away.

ICE DEVICE

Canadian engineer Reginald Fessenden designed a sonar system that could detect icebergs up to 3.2 km away, measure the sea's depth and send messages by Morse code. His invention was later fitted to submarines.

MAPPING THE SEABED

Today, sonar helps ships 'see' the seabed, even though it lies far below them in total darkness. They use it to study rock formations, and also to spot shipwrecks or sunken planes. This ship is using multibeam sonar, which sends out a fan-shaped signal. Echoes from the signal describe the shape of the sea floor.

PROXIMITY FUSE

Another wartime secret weapon used by the Allies was the proximity fuse. It was a kind of radar fitted into the tip of a shell fired from a gun. Radiation waves detected a target (such as a plane in the air), which made the shell explode.

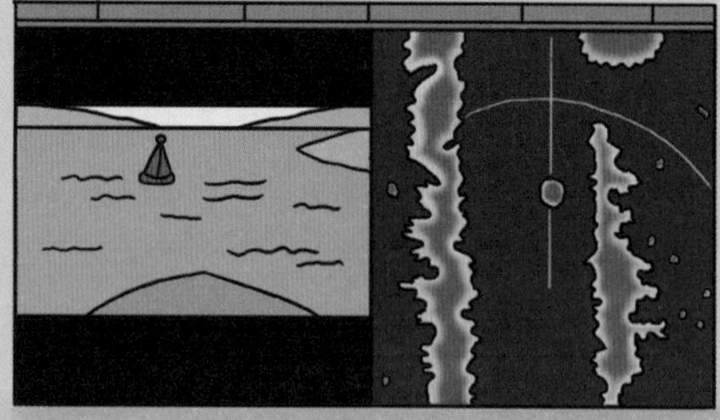

NAVIGATION AID

Today, radar is widely used for peaceful purposes, such as guiding ships at night. A ship's microwave radar scanner sends out radio waves and picks up reflections from solid structures, which show up on screen.

LASER LIGHT

If you've made it this far in the book, you're doing well. Let's face it, light is complicated stuff. And now we get to the really tricky subject: lasers! We use lasers every day – in CD and DVD players, for instance, and laser printers and laser displays. So who invented them, and how do they work?

MEET THE PHOTONS

To understand lasers, it helps to know a bit about light. Light is electromagnetic radiation – like a visible cousin of the invisible X-ray. In 1905, German scientist Albert Einstein described light as made up of tiny bits called photons, often shown as squiggles: wave-packets.

HOW A LASER WORKS

At the centre of a laser is a cylinder filled with a gain medium: a gas or crystal. At each end of the cylinder is a mirror; one with a small hole. The gain medium is 'pumped' with electricity or with light energy (photons). The photons excite the gain medium, which gives out more photons. Eventually, the photons fire out from the end of the laser.

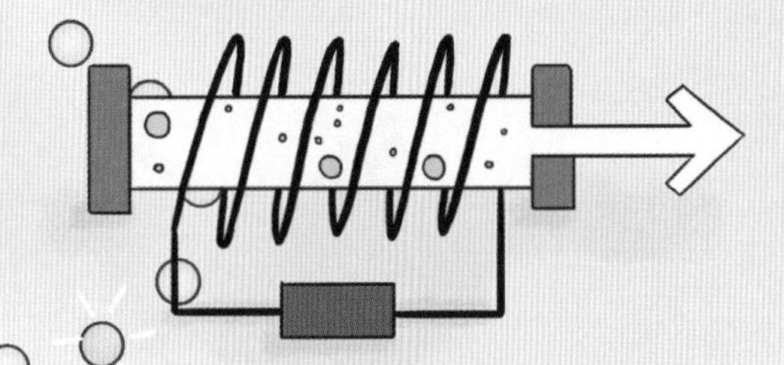

LASER BUILDERS

The first working laser was built in 1960 by Theodore Maiman in the United States. But he couldn't have done it without the pioneering work of other scientists in the US, Russia and Iran in the 1950s. Some of them are shown below.

My lidar is faster than your car...

SUPER-CUTTER

Lasers, once described as 'a solution looking for a problem', are used today in all sorts of devices. A cutting laser, for instance, will cut metal, wood and fabric. It can cut anything from very thin, delicate parts up to 12-mm-thick steel. The cutting pattern is computer-guided.

LIDAR GUN

Police use lidar guns to check on unsafe driving. Lidar works a bit like radar, but uses laser light instead of radio waves. It is super-accurate, homing in on a single car and gauging the speed in a split second. If you see one pointing at your car, tell your parents to slow down...

LASER BROOM

One of the dangers facing astronauts is the growing amount of space junk: bits of old spacecraft orbiting Earth at high speed. One speck of metal hitting your spacecraft could cause a catastrophe! One day, NASA may use 'laser brooms': Earth-based laser weapons that could zap the junk and push it out of harm's way.

CRAZY INVENTIONS

BOILED PEE

The grandfather of the glowstick, which is basically a chemical light, was German chemist Hennig Brand. In 1669, Hennig boiled a flask of his pee until it caught fire and turned into glowing phosphorus (the chemical that was later used in matches). Nice!

HORSE SENSE

The Dollond family of London were famous for inventing lenses for telescopes and spectacles, but in 1893 they went one step further when Mr Dollond invented bifocal glasses for horses. Why? The glasses made the road surface seem closer than it was, causing the horse to walk in a high-stepping manner (which was thought attractive in those days). Apparently, the horses enjoyed wearing them.

SPECS, NOT PECKS

It ain't just horses, either! In 1903, American Andrew Jackson invented tiny pairs of spectacles for farm chickens to wear. His aim was to stop them pecking each other, as this made them sick. The specs were primarily eye protectors and rose-tinted, to stop the hens becoming more frantic at the sight of blood, and pecking more. Jackson's idea worked...

LIEUTENANT PIGEON

Remember those aerial photographers? Well, in 1908, Dr Julius Neubronner of Germany invented a miniature camera... for pigeons. Strapped to the bird's chest, it worked by means of a timer, which went off regularly. He showed it off at public fairs, where people could buy postcards of pics taken by the birds. Later, Neubronner's pigeon camera was used in World War I. Crazy, but true.

LOOKING AROUND CORNERS

Hamblin glasses, invented in the 1930s, were special spectacles that reflected light at a 45° angle. They allowed you to lie flat on your back (if, say, you were unwell, or just tired) but still read from a book resting on your chest. If that sounds appealing, you're in luck: you can still buy them online!

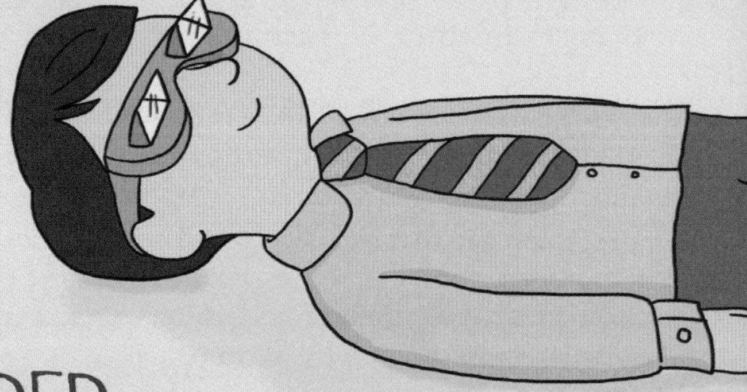

LIGHT-HEADED

Since the 19th century, solar baths had been a popular 'cure' for diseases of the head, such as a snotty nose or earache. This model is from the 1930s: patients put their head inside the big can, where an ultraviolet ray-gun fired artificial sunlight at it. UV light makes vitamin D, which is good for your body – but it also fries your skin and can cause cancers. Which is not so good.

ANTI-AIRCRAFT TUBAS

These things look like a monstrous brass band, but they're actually giant ear-trumpets. Armies used them in both world wars to listen for the engines of approaching enemy aircraft. When Robert Watson-Watt developed radar, war-tubas were no longer needed.

DAFT AND DEFT PREDICTIONS

THE ELECTRIC LIGHT BULB

Back in the 1870s–80s, when Thomas Edison was working on his light bulb idea, lots of people reckoned it would never work. Here are some of the daft things they said:

'Everyone acquainted with the subject will recognize it as a conspicuous failure.' –Henry Morton, president of the Stevens Institute of Technology

'When the Paris Exhibition closes, electric light will close with it and no more will be heard of it.' –Oxford professor Erasmus Wilson

'The electric light has no future.' –British inventor John Pepper

SWITCHED ON

Despite his gloomy (and obviously way wrong) prediction, Pepper was a huge fan of electricity. He used arc lamps to light up Trafalgar Square and St Paul's Cathedral in London for the wedding of Edward, Prince of Wales and Alexandra of Denmark in 1863. He also invented a way of projecting a 'ghost' of an actor onto the stage. It was so clever that even the brilliant scientist Michael Faraday begged him to explain how it was done.

Not so cleverly, when visiting Australia in the very dry summer of 1882, Pepper tried to make rain by firing big naval guns and rockets at the sky. It failed, and everyone laughed at him, so he gave up.

YES, IT DID COME TRUE...

Some people were better at making predictions. In his 1863 novel *Paris in the Twentieth Century*, French writer Jules Verne made several accurate predictions for the year 1960:

★ cities with electric lighting
★ skyscrapers
★ high-speed trains
★ a global message network (rather like the Internet)
★ fax machines
★ cars
★ the electric chair (yuk!)
★ weapons of mass destruction (boo!)

Amazingly, his publisher, Mr Hetzel, thought the story was too boring, so it was never published in Verne's lifetime.

SIMON SAYS

Astrologers study star charts and the zodiac to predict the future. They are sometimes wrong. Astronomers are very different: they are space scientists. But sometimes they, too, make mistakes.

In ancient Greece, Aristotle suggested the Sun circled the Earth, and later on, Ptolemy backed him up. We now know, of course, that it's the other way round: the planets in our solar system orbit the Sun.

More recently, in 1898, Canadian-American astronomer Simon Newcomb said: 'We are probably nearing the limit of all we can know about astronomy.' After a few more years' study, Simon wisely changed his mind, calling astronomy 'an illimitable field'.

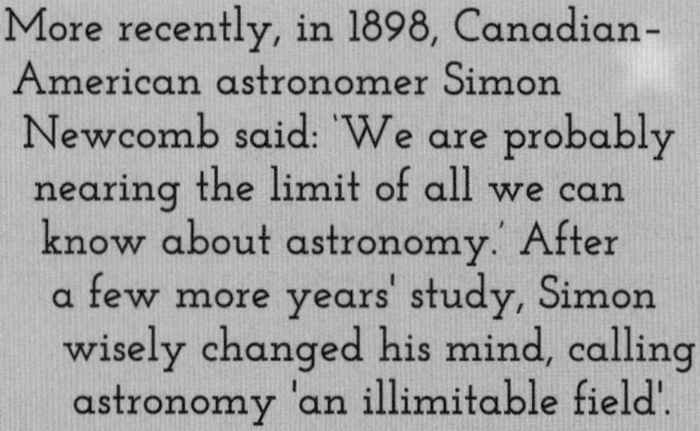

INDEX

The Author

British-born Matt Turner graduated from Loughborough College of Art in the 1980s, since which he has worked as a picture researcher, editor and writer. He has authored books on diverse topics including natural history, earth sciences and railways, as well as hundreds of articles for encyclopedias and partworks, covering everything from elephants to abstract art. He and his family currently live near Auckland, New Zealand, where he volunteers for the local Coastguard unit and dabbles in art and craft.

The Illustrator

Sarah Conner lives in the lovely English countryside, in a cute cottage with her dogs and a cat. She spends her days sketching and doodling the world around her. She has always been inspired by nature and it influences much of her work. Sarah formerly used pens and paint for her illustration, but in recent years has transferred her styles to the computer as it better suits today's industry. However, she still likes to get her watercolours out from time to time, and paint the flowers in her garden!